Tout le Monde peut cuisiner avec les BPF's!
Anyone can cook with the GMP's!

Spirulina laxissima

Sardinella aurita

By Dr. Mikel de Elguezabal M.
Biologist
A6 Director
A6-Solutions.com

Crypthecodinium cohnii

Ocimum basilicum

Undaria pinnatifida

Zingiber officinale

Saccharomyces boulardii

Arachis hypogaea

Fundación LEA Edit.

First Edition, 2015

Lactobacillus acidophilus

Agaricus bisporus

Cover and backcover: by Mikel de Elguezabal

First edition: 2015

© Mikel Alberto de Elguezabal Méndez, 2015

A6-Solutions.com, Uharte, Navarre, Spain

Editorial Fundación LEA, Calle Palmar, D-12, Riberas, 6101 Cumaná

Legal deposit: Public Libraries of Sucre State, Venezuela

ISBN 10: 1519249160

ISBN 13: 9781519249166

In a Virtuous Earth Colection

Printed in Ireland

Gemini International Printing.

Unit 4 Plato Business Park

Damastown, Dublin 15, Ireland

Index

Prologue……………………………………………………………………..1

Preface……………………………………………………………………2

Introduction...6

Introducing HACCP principles..9

Prerequisites programs & autocontrol systems: Good Manufacture Practices (GMP) and derivatives ……..19

Glossaries..90

Preface

Welcome to our first written contribution in Food Safety for food business owners and workers around the World, in which our company sets some basic goals to offer new education and pedagogy in this crucial matter for the humankind. Education is our first value as a company, cause we are all biologists from the academic and scientific sectors and we teach, educate and train every person in this World of Food Science, and we use the best ideas and most adequate and effective techniques to reach the level of learning that we want in every person taking our open or private courses on Hygienical Food Handling, Good Manufacturing Practices, HACCP, Food Microbiology, Thermic Treatments, Deep Frozen Food, etc.

We first bring the attention of our audiences by explaining what food is, and the source of our foods is, since always: Nature. We explain the composition of the cells of animal, plants or fungi we eat as food items, with the same molecules that compose our human cells and tissues and organs, to function and to give us life. Also the primary known enemies of our meals, the microorganisms, are living single cells or tiny multi cell organisms, both of microscopic features and ubiquitous in every, every grain of sand, soil, drop of water, rainwater, sweat, present in all human atmospheres, which are innocuous, the vast majority, harmful, often, and lethal, very few of the species. Bacteria, yeast, molds, microscopic protists, parasite cysts among others microbes well associated with different food classes and the environments where we achieve to cultivate crops, raise animals for food purposes or extract from the seas and rivers the fish harvest.

We rely in a majority of biologist partners in our existing franchises and future ones too, not only because the founder of this great team and dream was an inspired biologist who amaze everyday by the perfection and surprises of million species on Earth, mostly in his food microbiology labs during 30 years of teaching contents and values to food technicians in Cumaná, Venezuela, but also in the animals and plants he begun to study first in the Agronomy School and the ones he started to raise and seed in the familiar farm. We trust biology and biology related scientists and professors in our company because all food comes from nature, a majority of them come from life forms, and, inherited to all natural food there are vast numbers of microbes attached, in a wide diversity, but with a common basic cell function, molecules and biological laws that governs life, including the human body organs, tissues, cells and inside these, the organelles and its macromolecules (biomolecules): proteins, lipids, carbohydrates, all this floating in the cytoplasm of every biological species has its cell or cells, a cytoplasm made of water, salts, sugars, vitamins, organic acids, hormones, antioxidants, pigments and all the biochemical products and reactions (thousands!) happening each minute in life forms.

We mention protein, lipids, carbohydrates, vitamins, salts...that sounds like a food pyramid! that triangle that our school teachers teach us to get learned the lesson about healthy and balanced food consumption. These molecules of life forms (food) and nature (water & mineral salts) are key to our human body correct growth and development, where growing refers to the increase of cell size and numbers, increase of organs and organism size and adapting this to the proper development, attached to the quality of function, reached in a controlled maturation of cell and organs (organism) functions. Thus, growth= quantity (size, multiplication), and development= quality (type of functions, maturation). Both of these concepts are vital to all life forms, more ethically related to *Homo sapiens sapiens,* and its embryos, fetuses, babies, child, young, adults and elder persons. At each stage times of our lives, as seen before, we have different sizes, weights, metabolisms, energy, force, functions. But, always our bodies will need the rapid energy of carbohydrates for brain function and muscle movements, the energy store from lipids and the pigments, hormones and cell membranes that are constructed using natural lipids we eat, and the proteins to fabric our enzymatic machinery, related to all kinds of reactions in our cells, and to form the bricks of our muscle fibers, bone matrix, cartilages, cell membrane transporters and myriads more things of a major importance.

Thus, all food can provide us with healthy tools if these are composed of these natural molecules. But not the food made of inorganic colorants, flavors, edulcorants, odors, or super refined sugars, hydrogenated oils (margarines, *trans* fats). Food comes from different natural sources, as we observe in the front cover of this book: excellent sources of protein from algae, fungi, or even bacterial species like *Spirulina*, and excellent fatty acids from vegetables, herbivore fishes or even protists like *Crypthecodinium cohnii*, and vitamins, minerals and good carbohydrates from edible plant products like peanut, ginger, aromatic leaves, just as some of few examples of healthy food from biological species.

Health is secured if we eat properly and balanced enough as our school teachers told us, but the innocuous and safety/security of our food is related to perils that may occur during the consumption of not well heated meat, non well cleaned oysters, bad sanitary crop management, bad practices of hygiene at the food factory spaces, etc. These perils or dangers can be avoided by applying plans or programs at every level of food chains, from producers until processing and final consumers, if all actors involved in these food chains act professionally, rigorously, ethically without committing riskiness actions as not to wash our hands properly after the use of bathroom facilities to piss or defecate, both natural physiologic features of animal species like human beings are. Even presidents, pop stars, kings and queens have to evacuate fecal or urine wastes, daily, the most "democratic, egalitarian", of human features like that we have to eat good food, drink clean water or breathe clean air.

The main danger in food are human related actions, i) by its own hygiene, ii) by its own management of crops and raised animal, or contaminating with toxic modern human residues our soils, air and waters. The inner enemy for food microbiological contamination is our own body, full of skin pores, hair, epidermis cell falling each day, saliva, mucus generation in ears, nose, eyes, and of course the main perils of our bodies: genital and anal environments, full of most dangerous microbes for human food health. Maybe some bacteria or yeast does not harm our genital area or rectus, but, if these could reach our hands or skin, and then transfer to raw food or after heat treatment (post contamination of food preparations!) some dozen of bacterial cells (*Clostridium spp.*, for example) or a yeast cell (*Candida spp.*) could multiply there and we must develop lethal botulism or harmful oral candidiasis, among dozens of dangerous species that can spoil or contaminate our daily food intake.

Attitude, committing, rigueur, professionalism, maturity, are the main values that every (big or little) food business worker and owner must maintain in its daily actions at work, at home. The role of a food handler person is as crucial for public health as the role of a surgeon in a single patient, but with the difference that a mistake of a food handler would cost many persons getting ill (incidents) or worse, dead (accidents). The hands of a food handler must keep as properly as the hands of a surgeon, every minute of its labor.

Introduction

Food Production/exploitation in lands and seas, food Business, Industries or Companies related to the food economic sector worldwide, will be eternal activities as the Planet survives this new Millennium: yes, more than 7.000.000.000 of people has to eat and drink every day an estimate of 2000 kcal daily. However, some human beings do not meet this minimum, and others exceed it with the attached issues of health, wellbeing or ethical values that our societies develop in every corner of this, the unique Planet with *Homo sapiens sapiens.*

After the Food and Agriculture Organization, FAO , 1 out of 3 people worldwide are employed by agriculture activities, while just in Food industry like large food retailers or supermarkets, food-processing companies, or fast-food chains are employed more than 20 million people worldwide (ILO), but if we take in account the total of people that directly or indirectly works in the broad food sector of any country (gathering agriculture, processing, distribution, retail, restaurants, supermarkets, bakeries, and every small or big food business) we could reach around 50% of the Economically Active Population, which, *a grosso modo*, would represent 25% of the World population to produce, provide, transform, sell, cook and serve food for the entire population, including the food related workers themselves.

We argued that food sectors could be an "eternal" activity cause food and feed comes or derive from Nature in 100%, with higher proportions from organic or biological origin of food (from all kingdoms of Life forms! Edible animals, plants, fungi, bacteria and protists!). Apart of salts and water (mineral food?) for human, animal, or plant consumption, every living organism of these just mentioned, needs biomolecules to function, like proteins, lipids, vitamins or sugars, that compose our foods, be of natural fresh production or be of industrial candies or drinks, all this has (in a majority of cases, we hope) the organic or biological molecules that secures life in every cell and every human body, animal or life form. Why "eternal" thus? Well, like the Sun will not switch off in the next million years, and the photosynthesis carried out by plants, algae or phototrophic bacteria support the base of all biomolecules of the Biosphere: again, simple sugars, then polysaccharides, or amino acids to form proteins and enzymes, fatty acids to form lipids and liposoluble vitamins, hormones, organic acids to form hydrosoluble vitamins. These are the cell materials of the organisms we feed on, and are the cell materials we are composed of!, to carry out our metabolism, our lives.

As we see, all foods comes from nature, be organic or inorganic (mineral), all foods and the ingredients composing our daily meals or snacks derives from natural resources, harvested, fished, processed, transported, served, cooked, and finally consumed by each person in different cultural ways. All foods, thus, are in contact with the surrounding environments as the food chains and webs develops and could carry or accumulate biological, chemical and/or physical dangers and risks; risks that probably could be lowered or elevated in such degree if the society avoids bad practices, or not, in every step of our relationships with food production, processing and consumption.

The biggest problem right now is not only "how to feed" the larger World population, but, how to secure the safety, innocuousness and maintain health of consumers of this new food needed and produced, by avoiding microbial charges of pathogen bacteria, viruses, fungi or micro/macro parasites? How to reduce the presence of glass, metal or others non food particles in our dishes? How to reduce the exposition to hundreds of toxic chemicals in the air (David Suzuki *dixit*!), waters (surface waters, groundwater, rainwater, tap water!) and, of course, our aliments! These 3 types of dangers will be controlled to not happen in any of our legal food produced, processed and sold for human or animal consumption, as several official organisms and laws, directives, normative and even recommendations ask for all of us, actors in the scenario of food history. We would add, as an *addendum* to this book, another set of dangers that maybe are still not taken into account, but of authentic occurrence and may be related to a bunch of modern health issues: the "biochemical dangers". Or maybe in another book...

But before of "inventing" a new term to a real threat, we must develop the Food Safety programs that dictates FAO & WHO in the Codex Alimentarius (a conjoint Commission for Food, within the former UN organisms just mentioned), that is a mandatory reference in the food sector anywhere and for, the World food commerce, distributors, producers and traders. We must educate people related to food, from producers, processors and consumers, on how to identify and then avoid the biological, chemical and physical dangers in foods, and how to manage without risks or avoiding riskiness actions during every step of any existing or future food chain for any product.

Actually, we want our worldwide readers have the information about what programs are available around the World since XX century, like food based GMP and HACCP or the International Organization for Standardization (ISO) that deals with the food safety in the ISO 22000, an international standard program derivative of ISO 9000, that specifies the requirements for the food safety management systems that must follow the following elements:

- interactive communication
- system management
- prerequisite programs
- HACCP principles

As ISO is developing additional standards that are related to ISO 22000, these are known as the ISO 22000 family of standards:

ISO 22000 - Food safety management systems - Requirements for any organization in the food chain.

ISO 22001 - Guidelines on the application of ISO 9001:2000 for the food and drink industry (replacing: ISO 15161:2001).

ISO/TS 22002- Prerequisite programs on food safety—Part 1: Food manufacturing

ISO TS 22003 - Food safety management systems for bodies providing audit and certification of food safety management systems.

ISO TS 22004 - Food safety management systems - Guidance on the application of ISO 22000:2005.

ISO 22005 - Traceability in the feed and food chain - General principles and basic requirements for system design and implementation.

ISO 22006 - Quality management systems - Guidance on the application of ISO 9002:2000 for crop production.

ISO 22000 is also used in the Food Safety Systems Certification (FSSC) Scheme FS22000. FS22000 is a Global Food Safety Initiative (GFSI) approved scheme.

Because these ISO programs are related to bigger food business that have secured their products properly, and, our objective with this document is to give every person involved in food production, processing and expending own business, a chance to ameliorate their practices with our sacred aliments, developing good, better each day and best food manufacturing and the highest hygiene-safety practices for our breakfast, lunch, dinner, supper or snack meals, we will construct this document as follows: i) introducing briefly the HACCP principles, ii) applying the prerequisite programs or autocontrol systems, that are included inside the Good Manufacture Practices (GMP), Standard Operation Procedures (SOP's) and Standard Sanitization Operation Procedures (SSOP's) and finally, the base of all programs above, iii) educating ourselves in the highly important Hygienical Food Handling, adjusted to the size of medium, smaller and microbusinesses around the planet Earth, the vast majority of little buildings and farms dedicated to food production, processing and restoration or derived food vending businesses. Let's go to learn on Food Safety's best plan!

1. Introducing the HACCP principles

There are the Seven Principles of *Hazard Analysis and Critical Control Point* (HACCP) System. In order to enhance food safety, every stage of the food production (from purchasing, receiving, transportation, storage, preparation, handling, cooking to serving) should be carried out and monitored scrupulously. The HACCP system is a methodical, scientific, rigorous, continued and systematic approach to identify, assess and control hazards in the food production processes. With the HACCP system, <u>food safety control is integrated into the design of the process</u> rather than relying on end-product testing. Therefore HACCP system provides a <u>preventive</u> and thus cost-effective approach in food safety.

The seven principles of a HACCP System are:

1. Analyze hazards

Food safety hazards are biological, chemical or physical properties that may cause a food to be unsafe for human or animal consumption. We have to train and educate in such way that we could analyze, view in perspective, focus minded, to achieve pointing the potential hazards of your food business in order to identify any (all) hazardous biological, chemical, or physical property in raw materials and processing steps, and to assess their likeliness of occurrence and potential to render food unsafe for consumption of your clients at your bakery, restaurant, farm, or artisan food fabric!. Print the next table and do the work team needed to achieve this first goal.

Hazard Analysis Sheet

Steps of the Process					
Processing Point	Possible Hazards: 1-Biological 2-Chemical 3-Physical	There are hazards of significance to food safety of your product?	Decision arguments?	Action for Prevention?	Critical Control Point? Y/N

Food Business Name:_____ Characteristics of the Product:_____ Address:_____
Date:_____ Signature:_____

2. Determine critical control points

A critical control point is, "a point", a step or a procedure in a food manufacturing process at which control can be applied and, as a result, a food safety hazard can be prevented, eliminated, or reduced to an acceptable level. Not every point identified with hazards and preventive measures will become a critical control point. A logical decision-making process is applied to determine whether or not the process is a critical control point. The logical decision-making process for determining critical control points may include factors such as:

-whether control at this particular step is necessary for safety;

-whether control at this step eliminates or reduces the likely occurrence of the hazard to an acceptable level;

-whether contamination with the hazard identified could occur in excess of acceptable levels;

-whether subsequent steps will eliminate or acceptably reduce the hazard.

Print the next table and do the team work needed to achieve this second goal.

Resume of the HACCP Plan Sheet

Critical Control Points (CCP's)								
CCP's	Hazard Description	Critical Limits in each Control Action	Monitorize				Corrective Action?	Archive Records of Actions/Procedures
			What	How	Freq.	Who		

Food Business Name:_____ Charactheristics of the Product:_____ Address:_____ Date:_____
Signature:_____

3. Establish limits for critical control points

Limit for critical control point is a criterion which separates acceptability from unacceptability. It is the maximum or minimum value to which a physical, biological, or chemical hazard must be controlled at a critical control point to prevent, eliminate, or reduce to an acceptable level the occurrence of the identified food safety hazard. Examples of limits for critical control point are time, temperature, humidity, water activity and pH value. The limits should be measurable. In some cases, more than one critical limit is needed to control a particular hazard.

Cold chain units Sheet

Cold temperature Registry (Refrigerators, Cold Chambers, Counters, Freezer tunnels)								
Location & Description of Unit	Date	Time	Temp. (F°/C°)	Corrective Action	Food Handler Operator Name	Supervisor Name Date/Time	Manager Name supervising	Observations

How to act: the responsible food operator employee must write down and record the location and type of Cold Unit, observing data as air temperature, integrity of the unit, and cleanness of the unit surfaces. Date, time (using military format), corrective actions and names of all members of the food business involved must be recorded on this sheet. Supervisors and managers have to verify *in situ in vivo*, by random monitoring, these actions of responsible food operator employees. This sheet must be archived at least by 3 months after selling the product if you have a Restaurant, Butchery or related to Cold chain foods, or up to the end of the shelf life denoted in the product label information, if you manage a food industry for canned food, bottled drinks, packed food or other non perishable food.

4. Establish monitoring procedures for critical control points

To monitor a process is to plan a continuous sequence of observations and measurements to assess if a critical control point is under total control and to produce a rigorous record for future uses in HACCP: the Verification!. Monitoring is very important for a HACCP system. Monitoring can tell the workers, supervisors and managers of any food business, little or big, if there is a pattern or trend towards a loss of control so that it must be taken into immediate action to bring the process back into control before the limit is exceeded.

The person responsible for this HACCP monitoring procedure is 100% committed in this basic and key element of the plan, and it is trained properly in the operations inherited to the raw food, processed food, equipments, machines, devices or area of the process, he or she is "monitoring", focused enough to know the important role for the safety of the food produced in its company (where he/she works or which he/she owns).

Calibration Thermometers Monitoring Sheet

Calibration of Thermometers							
Date	Time	Thermometer Number ID	Reading of reference Thermometer	Reading of Thermometer under calibration	Requires adjustments (Y/N)	Corrective action	Name of the responsible

Verified by:_____ Date:_____ (To archives)

How to act: the responsible food operator has to calibrate these maximal importance instruments, daily, if possible each turn of activities in the food business (twice a day), and correct if necessary, any deviation with the confidence given by the reference thermometer (which in turn, must receive periodical "third party" calibration). Supervising by visual observations of these daily calibrations made for the responsible person is needed to verify the correctness and on time achievement of this HACCP step. The supervisor of this operation must sign this sheet each time of review procedures, with the document maintained for 3 months, archived, after selling the product if you have a Restaurant, Butchery or related to Cold chain foods, or up to the end of the shelf life denoted in the product label information, if you manage a food industry for canned food, bottled drinks, packed food or another non perishable food.

5. Establish corrective actions

Corrective action is an action taken when the results of monitoring at the critical control point indicate that the limit is exceeded, i.e. a loss of control.

Since HACCP is a preventive system to correct problems before they affect food safety, plant management has to plan in advance to correct potential deviations from established critical limits. Whenever a limit for critical control point is exceeded, the plant will need to take corrective actions immediately.

The plant management has to determine the corrective action in advance. The employees monitoring the critical control point should understand this process and be trained to perform the appropriate corrective actions.

Corrective Actions Sheet

Corrective Actions		
Responsible person in charge:		
Product:	Lot or Batch code:	Observations
Deviation:		
Cause:		
Cause of Deviation controlled by (Name of responsible person)		
CCP Controlled after corrective actions?		
Action for future prevention:		
Is the Product disposed to commercialize?		

6. Establish verification procedures

Verification is the application of methods, procedures, tests and other evaluations, in addition to monitoring, to determine compliance with the HACCP plan.

Some examples of verification are the accurate calibration of process monitoring instruments at specified intervals, direct observation of monitoring activities, and corrective actions. Besides, sampling of product, monitoring archived record reviews and inspections can serve to verify the HACCP system.

The food business owner or manager should check that the employees are keeping accurate and timely HACCP records.

Heat Control Sheet

Cooking/ Heat/ Reheating temperatures						
Date and Time	Food item/ Heat process	Internal temperature and time of treatment (Twice)	External temperature	Corrective Action taken	Name of responsible person	Supervised by Supervisor/ Manager

How to act: the responsible person has to write down and register the product or preparation name, the double temperature readings and the time of heat treatments, and corrective action taken. The supervisors have to verify, visually *in vivo*, that responsible persons have accomplished these temperature controls. This sheet must be archived at least by 3 months after selling the product if you have a Restaurant, or related to *in situ* cooked foods, or up to the end of the shelf life denoted in the product label information, if you manage a food industry for canned food, bottled drinks, packed food or another non perishable food which uses one or various heat treatment processes.

7. Establish a record system

Maintaining proper HACCP records for each step or process of those mentioned before, but also for the next examples or new control sheets created by you, is an essential part of the HACCP system. Accurate and complete HACCP records can be very helpful for:

-documentation of the establishment's compliance with its HACCP plan;

-tracing the history of an ingredient, in-process operations, or a finished product, when a problem arises;

-identifying trends in a particular operation that could result in a deviation if not corrected;

-identifying and narrowing a product recall.

The record of a HACCP system should include records for critical control points, establishments of limits, corrective actions, results of verification activities, and the HACCP plan including hazard analysis.

To establish recordkeeping procedures, plant management may:

Develop forms to fully record corrective actions taken when deviations occur; identify employees responsible for entering monitoring data into the records and ensure that they understand their roles and responsibilities

Holding units Sheet

Holding units times and temperatures									
Date	Food item	1° Measurement Entrance			2° Measurement Exit				Corrective Action
		Time	Temp.	Name of responsible person	Time	Temp.	Name of responsible person		

Supervisor name:_____ Date:_____

How to act: Responsible persons in this role must write down and record the temperatures when any food lot or batch enter and exit the holding.

Additional Control Sheets

Receiving and Transporting Sheet

Reception / Transportation							
Date	Time	Supplier	Product name	Temperature (°C)	Corrective actions taken	Name of responsible person	Name of supervisor/ manager. Date

Hot to act: Responsible person and (visually verified) supervisors/managers must secure the records on this sheet of temperatures and corrective actions for all delivered and received food commodities or processed/sold food. Monitor each shift of responsible employees and review this sheet at every turn/day of work of your company. This sheet <u>must be archived</u> at least by 3 months after selling the product if you have a Restaurant, or related to *in situ* cooked foods, or up to the end of the shelf life denoted in the product label information, if you manage a food industry for canned food, bottled drinks, packed food or another non perishable food which uses one or various heat treatment processes.

Cooling temperatures Sheet

Cooling temperatures control Cold Unit name and capacity: _____									
Date	Food item	Time/ temp	T/t	T/t	T/t	T/t	T/t	Corrective Action taken	Name of responsible person	Name of supervisor/ manager, Date

How to act: the person responsible of this role must keep the records of temperature readings each hour during each work turn, as well as possible corrective actions taken, indicating is the cooling unit is Empty, half full, 3/4 full or overloaded. Supervisors and managers must accede often to the cooling unit to verify visually the proper work of the responsible persons, and write down their names, dates and hours. This sheet <u>must be archived</u> at least by 3 months after selling the product if you have a Restaurant, or related to *in situ* foods, or up to the end of the shelf life denoted in the product label information, if you manage a food industry for frozen food, canned food, bottled drinks, packed food or another food product which uses one or various cooling treatment processes.

Cleaning + Sanitizing of food contact surfaces Sheet

Food contact surfaces cleaning & sanitizing control								
Date/time	Wash temp.	Rinse temp.	Final rinse temp. during sanitization	Attach Heat sensitive tape (for example, from *Getinge* consumables) here	Sanitizer concentration, ppm	Corrective actions	Name of responsible person	Supervisor / manager name, Date

How to act: responsible persons must maintain a record of times, temperatures of water and surfaces, concentration of sanitizers and corrective actions. Also the supervisors and managers must verify visually, in vivo, that the responsible persons are achieving the correct procedures of cleaning and sanitizing at any turn shift, writing down their names and dates each day/turn. This sheet <u>must be archived</u> at least by 3 months after selling the product if you have a Restaurant, or related to *in situ* foods, or up to the end of the shelf life denoted in the product label information, if you manage a food industry for frozen food, canned food, bottled drinks, packed food or another food product in facilities subject to regular cleaning and sanitizing.

Food preparations Sheet

Food preparation controls								
Date/ start time	Product name	Temp'	Temp"	Prepared food Weight	Corrective actions	Final time	Name of responsible person	Name of supervisor/ Managers/date.

3 important concepts to deal in HACCP:

Monitoring: is the process of checking to ensure a pre-determined set of standards or requirements are met: for example, monitoring the internal temperature of the product to ensure it reaches a set point determined in your HACCP System.

Verification: is the process of ensuring the monitor is following the correct procedures and all requirements are met. Record verification would require a different person than the monitor to review the records to ensure all limits were met and the records are complete. On-site verification is when someone observes the monitor completing the task to ensure they are following the correct procedures as outlined in the written program.

Validation: is the process of proving that what is being monitored is controlling the hazard. For example, you may validate that cooking to 71°C will kill *E.coli 0157:H7* in your product.

A very good example of an HACCP system for a family own food business, is this courtesy of Fletchers Bakery, Sheffield, England, which is available on the web:

http://be-sy.org.uk/pdfs/PDF%20Science/Fletchers/2.5%20FletchersHACCPanalysis.pdf

2. Prerequisites programs & auto control systems: Good Manufacture Practices (GMP) and derivatives*.

The <u>Good Manufacturing Practices</u> (GMP's) and associated programs are basic sanitary and processing requirements needed to ensure production of healthy & safety food. This Plan, as any other existing or to be developed in the future, is <u>like a game instructions or a culinary recipe</u>. Let's see, there are many rules to accomplish and there <u>are pieces of a puzzle</u> to execute in GMP:

1°, GMP's for Building & Facilities

2°, GMP's for Equipments & Utensils/"Food Tools"

3° GMP's for Good Quality Water supply

4° GMP's for Personnel (Hygiene Controls/Continued Education)

5° GMP's for Hygienical Requisites for Food Production (*SOP's)

6° GMP's for Sanitization Programs (*SSOP's)

7° GMP's for Storing & Transport

8° GMP's for Bathrooms, Restrooms/Toilets

9° GMP's for Residues Management

10° GMP's for Plagues/Pest Management

11° GMP's for Chemical Products Control

12° GMP's for Glass and Foreign Material Control

13° GMP's for Food Defense/Prevention of Intentional Food Contamination

14° GMP's for Allergens Ingredients Control

15° GMP's for Clients Complaints

16° GMP's for Proper Labeling & Traceability/Documenting/Archives

Figure 1. Good Manufacturing Practices puzzle, as resumed in this A6 document for educational purposes of its clients.

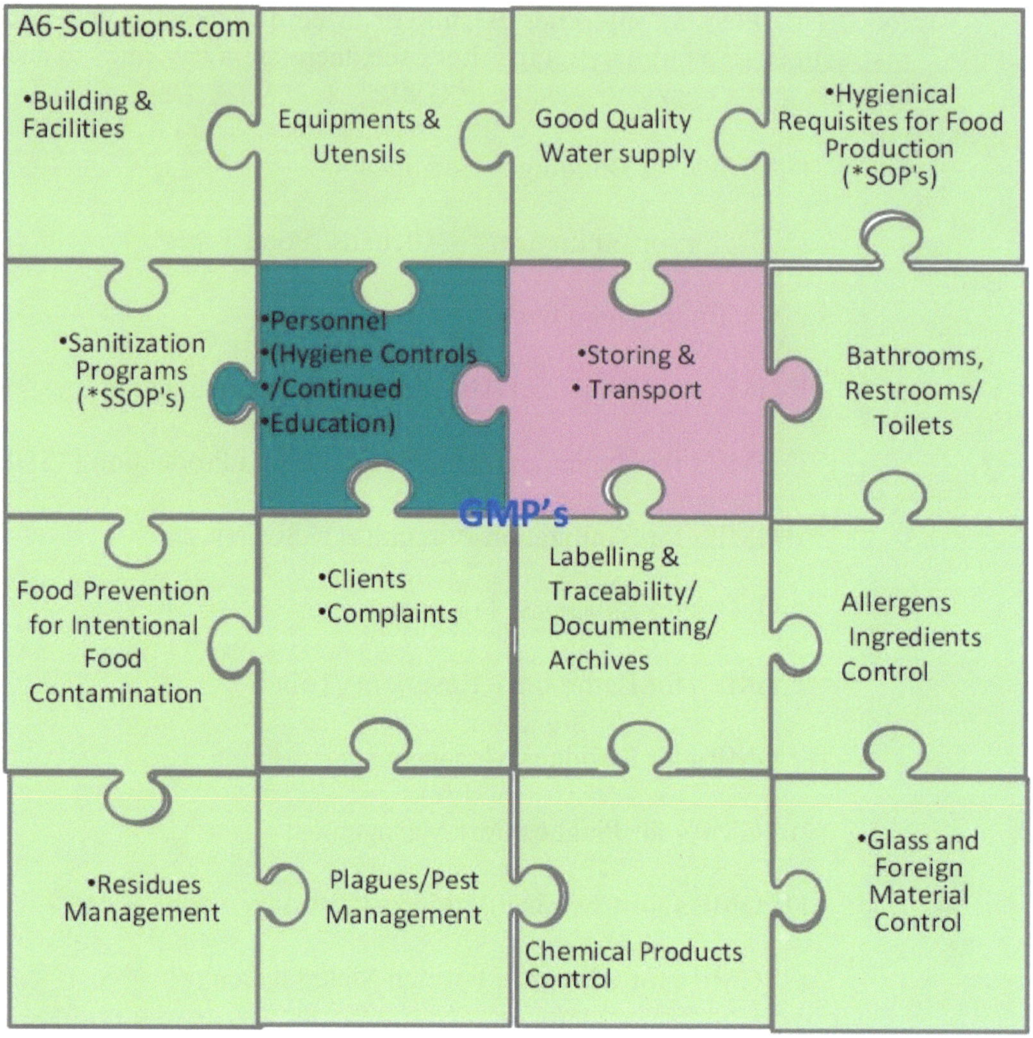

1°, GMP's for Building & Facilities

Good Manufacturing Practices. Buildings and Facilities
*
GMP plan requires that our facilities must be designed and constructed in such way that floors, walls, ceilings and all architectural features of the food business do not contaminate food (raw or prepared food, all ingredients, any additives), or all surfaces with potential or planned contact with food, or food packaging materials.

Floors, walls, ceilings, and doors in all areas where food is handled or stored must not absorb, never, water or any liquid substance, and these should be of the easier design to keep it clean and to facilitate sanitization. The selected materials should maintain integrity during a long "life of use", like stainless steel, and not be susceptible to cracks, scars, or pitting when we apply the soap, detergents, sanitizers, and brushing, mopping and in the normal conditions of use planned to this construction.

The corners, that are areas where the floor and two walls meet must be covered by a waterproof seal to avoid accumulation of water, chemicals and food debris in that sensitive area that could act as a microbial niche in your food business. Every little crack in floors or walls would be an excellent place where bacteria, mold, yeast or algae (microorganisms you will not see, but its colored cotton like spots) can grow and potentially reach your raw or prepared food!. A tiny hole in the floor would attract pest rodents, birds and insects if it is full of water and "leftovers"; this is basic, logical, so, just ask yourself if your floors, walls and ceilings are constructed with durable, non corrosive, non absorbent materials that are easy to keep clean and sanitized and be always in a good repair state, thus obeying the "law" of this GMP step?

The maintenance of facilities.
The outside of your food business or plant needs to be maintained with the aim of discouraging pests. The facilities and its structures shall be suitable in size, construction, and design to facilitate maintenance and the sanitary operations for food manufacturing purposes. The worker/owner of the food business, anywhere in the World, must keep attention on this list of GMP requirements in the matter: • Floors, • Walls, • Ceilings, Ventilation and • Lighting.

All these areas must be made of <u>acceptable materials</u>, avoiding some dangerous (to food, to your clients, for your business!) materials. You have to supply your facilities with food processing equipments and architectural features within a range of materials that meets the safety for food about the chemical, physical and biological dangers we talk earlier. There are materials for construction and for surface covering that are designed just for food business and plants, like <u>stainless steel</u> tables (fixed or rolling), equipments, etc. <u>We discourage the use of aluminium</u> by its possible links to Alzheimer, Parkinson and other health problems via contamination of food by slow but existing corrosion, also Copper or Iron metallic surfaces would contaminate your food slowly and sick your clients, and your business. We would use <u>Copper for non food</u>

contact surfaces, for example, in a future use as a "wallpaper" material to cover without any pore and crevice the walls of every room in your food business; Copper has an antimicrobial activity of the current demonstration in microbiology labs lessons worldwide by its ionizing energy that prevents the growth of bacteria and fungi.

We also encourage our courses, students and future entrepreneurs in food businesses, that if possible, use Silica materials for working surfaces in food preparing, like hard glass (smooth and well polished), ceramic without rough joint surfaces. Silica made material are the most secure for food in the case of preventing chemical contamination, cause it is an element (the most abundant in the Earth cortex!) innocuous, inert, safe, also for preventing the biofilm formation ("slime like" microbial visible spots) if the smoothness is secured daily and the cleaning + sanitization is carried out properly, although Silica just would represent a risk if broken into pieces of any size of glass particles, that are tremendous perils to food.

Options like Titanium, Platinum or Gold for food contact surfaces appears too expensive for the majority of food business and restaurants we are directing this educative document.

Evidently, organic materials, full of Carbon molecules, be of synthetic origin like plastics of any kind, or be of natural origin like wood, have many chance of develops cracks, crevices, have too many pores, that are absorbent not only for water, but for many liquid chemicals like many pesticides, lubricants, greases, oils, fuels...and, also those pores and crevices are the perfect niches for microbial growth with them moist ambient, the food particles and the necessary time for bacteria or fungi to multiply in vast numbers! For last, these materials (wood and plastics) are degradable and its particles or debris could reach the food your preparing today. It is risky to choose these materials to your food business.

Some kitchens around the planet uses working surfaces made of granite or marble rock, with microscopic pores and cracks, that absorb water, oils and every liquid substance, and would permit the growth of molds, yeast, algae and/or bacteria. Yes, these materials are beautiful, and if you clean it daily, well-done, but better for houses than for public food business, where, with the amount of work you have prepared food ready to eat or to distribute later, that volume of food activity elevates the probabilities of these surface contamination, cause these are capable of develop water or liquid accumulation and be a microbe harbor to reach your food.

So, a sanitary design in all business for its food contact surfaces must be smooth; impervious; free of cracks and crevices; nonporous; nonabsorbent; non-contaminating; nonreactive; corrosion resistant; durable and maintenance free; nontoxic; and cleanable.

Stainless steel rolling table.

Copper industrial foils, an A6 proposal "smooth wall cover" for the future.

Ceramic working surfaces with porous joints in a house's kitchen, not recommended for food businesses

A hard glass surface example appears cleanable and sanitizable for a restaurant's kitchen.

Some GMP sets as mandatory the unacceptable conditions or materials for doors, floors or walls that will eventually cause food products to become contaminated. Damaged floors, wall panels, ceilings or wooden structures can provide a place for pests to hide, build a nest, breed, and eventually infest your entire building. These damaged areas need to be replaced or repaired the day they are discovered.

To meet the GMP requirement that floors, walls and ceilings be in good repair and kept adequately clean to prevent contamination of food, food contact surfaces or packaging materials your food business (you) will need to: i) Evaluate the materials and the condition of all floors, walls and ceilings; ii) Repair or replace any areas that are not made of suitable materials, which are damaged, or are otherwise unable to be kept clean and sanitary, iii) Monitor the condition of floors, walls and ceilings routinely so that damaged or unsuitable conditions can be repaired before contamination occurs.

The food business or plant and its facilities shall be constructed in a manner that prevents dripping water or condensate from overhead fixtures, ducts and pipes, and coolers from contaminating food, food contact surfaces or packaging materials. Food business facilities shall provide adequate ventilation or control equipment to minimize odors and vapors (including steam and noxious fumes) in areas where they may contaminate food; and locate and operate fans and other air-blowing equipment in a manner that minimizes the potential for contaminating food, food-packaging materials, and food-contact surfaces. Improper ventilation in the food plant can also be a source of contamination. Your business will need to evaluate the air, air quality, and movement of air in your facilities. The systems you have for maintaining air temperature and humidity also needs to be maintained, monitored and recorded (archived).

Food processing, handling or storage facilities needs to have some type of ventilation system that prevents the build-up of heat, steam, condensation or dust that could contaminate food products. A properly designed, installed and operated Heating, Ventilation and Air Conditioning systems are needed to maintain the proper conditions that protect both the food you offer and the persons working in your food business (maybe yourself and your family).

The fans or air blowing equipment should be located and operated in a way that minimizes the potential for contamination of food, food contact surfaces, and food packaging materials. Monthly, please, take a walking survey around your facilities. Using your eyes and smell, check for condensate, steam, dirty fans and odors.

The food business facilities must be provided with adequate lighting in handwashing areas, dressing and locker rooms, toilet rooms, and in all areas where food is examined, processed or stored and where equipment or utensils are cleaned; and provided with plastic or acrylic protectors for safety-type light bulbs, fixtures, skylights or other glass suspended over exposed food in any step of preparation, protecting food against physical contamination in case of glass breakage by the explosion of bulbs, for example. The providing of an adequate light level in food handling, processing or storage facilities need to have enough to ensure that the areas are being easily monitored by our own eyes or supervisor's eyes to verify the proper order, the right cleanness, the best sanitization procedures. Adequate light is necessary in every area of the facility, including areas where food is processed, handled or stored, as well as locker rooms, hand wash areas, and toilet facilities. Your food business can prevent glass contamination and the risks associated with the glass in bulb lights possible broking and contaminating of food products (in any step: raw or ready to eat!) must be considered as lethal and prevented rigorously. Please, repair, replace or modify any light fixtures that do not provide adequate light or are not adequately shielded or made of shatter resistant material. Develop a routine procedure for the inspection deliveries of replacement light bulbs, fixtures, or safety shields to make sure that they accomplish the safety requirements.

How to Monitor

A routine daily visual inspection of light sources in all areas of the food business must be made. Check that all replacements meet the requirements set for proper shielding or bulb type.

Non protected light bulb, Forbidden for food business!

An acrylic protector option for ceilings light bulbs.

Facility Design

Your food business or plant and its facilities shall be constructed in such a manner that aisles or working spaces are provided between equipment and walls and are adequately unobstructed and of adequate width to permit employees to perform their duties and to protect against contaminating food or food-contact surfaces with clothing or personal contact.

About the adequate space, the conditions in your food business need to be evaluated to be sure that there is enough space to do your work. Food processing must be done in a space that is large enough for employees and the equipment that they are using to move around without contaminating food. This may <u>require planning and organization to design a workspace with workstations</u> that have adequate space: a Plant Layout! or the diagram where the logical flow of a production area might be designed to provide adequate space and a linear-one-way movement of products and ingredients, additives through your processing area or restaurant's kitchen. When processing or preparing food, the <u>best layout provides maximum separation of raw materials from finished products</u>. Ideally the <u>raw material enters the plant at one end</u> of the facility and moves through each step in the process until the <u>finished products are packaged at the other end of the facility</u>. The movement of people and equipment between processing areas should be managed so that contamination from raw product areas is not transported to more sensitive areas where finished products are handled, cooked, sealed or stored. We propose you, dear food business owner, as a require, that you develop a floor plan or diagram, with your experience for any and all products, dishes and food you create, by planning, thinking in the necessary steps (in the right order, logically, as a technical recipe for safety foods) to create adequate workspaces for the tasks you are doing. The GMP regulation does require that food be protected from contamination that could be caused by a work environment that is too crowded or congested. To meet <u>this requirement you should periodically assess the layout of each processing area to assure that processing procedures are being followed as designed for the processing area and that changes in products do not lead to potential sources of contamination</u> because they were not originally included in the plans when the facility. To meet this GMP requirement and provide adequate aisles or working spaces between equipment and walls that are unobstructed to permit employees to do their work and prevent contamination of food, food contact surfaces, and packaging materials you will need: i) <u>to evaluate</u> the layout and placement of equipment, processing lines, and work stations to make sure that aisles and workspaces are adequate and unobstructed and designed to prevent contamination; ii) <u>to relocate</u>, move, or otherwise modify placement of equipment, processing lines, or workstations so that adequate space is provided for employees to perform their duties and contamination is prevented; iii) <u>to monitor</u> routinely all areas where food, food contact surfaces and packaging materials are used to make sure that adequate, unobstructed space is provided.

Periodically, the owner or a responsible worker should observe the activities conducted in all areas of the plant at several different times of the day and monitor the impact of any design, structural and material changes that could exist. We still encourage you, as a requirement, to monitor archived records, because you must keep records of the results of your observations for your own use, without expecting the official visit, being methodical in advance, gaining time and efforts. If any actions are necessary to correct the existing problems, then these actions must be noted on the written record.

If you have it, the surrounding grounds of your food business or plant must be free of weeds, and totally covered by structured material that can support heavy duty machinery and vehicles (asphalt, concrete, cement), and never be of gravel, sand, rubble or "nude soil" where could grow weed plants that converts in the harbor for microbes and feed for pests. This important construction in your food company outsides, must keep almost the same performance as is was designed to be in contact with food, ingredients, additives or surfaces and materials that have contact with the food product; this is: as an indoor floor of the food business, without water accumulation, without cracks or crevices that accumulate residue, and with a slope inclined towards the drainage of the out borders of the food company, never directing the rain water and wash water with residues towards the entrance and interior drainages of the food plant.
• Facilities – cleanable surfaces, good ventilation and lighting, chemical control.
• Well-maintained hand washing and restroom facilities.

<u>Develop a Monitor plan for this GMP piece of the puzzle and each of its components.</u>
Conduct a <u>visual inspection</u> of the condition of all areas of the plant daily. Although the current GMP does not require monitoring records, we implore you to keep a record of the results of your observations for your own use. Design your own GMP sheets. If any actions are necessary to correct problems, these actions should also be noted on the written record. *That is being and behaves in a rigorously and seriously scientific like attitude!*

2°, GMP's for Equipments & Utensils/"Food Tools"

The equipments, tools and utensils related to food processing or those to have contact with food in any step of your layout, must be of a proper sanitary design and constructed of adequately cleanable materials, food-grade, suitable for intended use, precluding food contamination •

So, the equipments have to be installed and maintained in such way to facilitate cleaning of equipment and the adjacent areas. You can not use, but only food-grade lubricants in your machines, and clean + sanitize it properly

All the types of equipments your food business will need in order to receive, process, or store food, must be maintained safe, secured, in a way that will prevent it from getting contaminated from chemical, physical and/or biological dangers.

The "proper design" concept rely on material of construction of the machine (equipment) or tool, and on the form/shape and features of a safe machine for food processing purposes. There are fabrics around the planet dedicated to construct and sell food business machines under another GMP program, this specialized in food equipment GMP, as there are GMP norms in each country that dictates companies producing cans, bottles, bags for food products. You will contact these companies to secure your food business.

Thus, think of a proper design, construction, and use of food processing equipment; in the design, construction and the use of utensils and food contact surfaces; think in a proper design and construction of manufacturing systems; in monitoring devices to measure temperature, acidity, water activity or other conditions necessary to prevent the growth of undesirable microorganisms in the food you handle process or store.

All of the equipments and tools that you use in your food business and processing operations must be made in a way that will not contaminate the food (not Aluminium, Iron, bad plastics, wood). This means that equipment and utensils must be made of certain materials (as seen in the anterior section), and be fabricated, installed and maintained properly.

The GMP program around the planet coincides in demanding these requirements: i) All plant equipment and utensils shall be so designed and of such material and workmanship as to be adequately cleanable and shall be properly maintained; ii) The design, construction, and use of equipment and utensils shall preclude adulteration of food with lubricants, fuel, metal fragments, contaminated water or any other contamination; iii) Food contact surfaces shall be corrosion resistant when in contact with food, made of nontoxic materials, and designed to withstand the environment of their intended use and the action of food and cleaning compounds and sanitizing agents; iv) Food contact surfaces shall be maintained to protect food from being contaminated by any source, including unlawful food additives; v) Seams on food contact surfaces shall be smoothly bonded or maintained so as to minimize accumulation of food particles, dirt and organic matter and thus minimize the opportunity for growth of microorganisms.

Think about this if you want your food business's "machines & tools" to display the GMP prevention systems and protect your brand and products: the type of material used, its design and construction, its maintenance and durability, and the ability to be easily and adequately clean and sanitize it.

*Type of Material: Stainless steel is the preferred material for wet, acidic foods, milk, fish, shellfish, oily or fatty foods, and for meat, poultry, and all food. Stainless steel is durable and can be fabricated with a smooth, cleanable finish. These equipments are expensive, just once, but they last for decades if you secure a good use and a minimum maintenance. We have clients with 70 years old stainless steel tanks in a brewery that looks like brand new today!

Although there are a variety of plastic tools, and "each week" there is a new plastic polymer invented, we discourage these investments for your food business: they broke, they could be a source of chemical contamination of food, and they could harbor microbes "eating" plastic organic molecules made full of Carbon, forming the dangerous biofilms in everywhere.

Equipment and Utensil Design and Construction: Food processing equipment and food contact surfaces should be designed and constructed so that they can be easily cleaned. For equipments this means that it can be separated easily and that all its individual parts can be cleaned and sanitized. About the "dead spaces" the equipment should be designed to avoid sharp angles, structures, or dead spaces that can collect dirt, food debris and water. These areas are difficult to clean and can provide a hiding niche for pathogenic microorganisms that could contaminate your food products and the entire business local. Equally the seams and joints should be smoothly bonded; the metal joints should be welded with a continuous weld, which is ground to a smooth finish. Avoid crevices, cracks, weld debris and burns, these must be removed to make cleaning easier and the sanitization. The solder should only be used if necessary, and solder and flux must not contain any toxic metals such as lead, cadmium and antimony. The bearings and gears of all machines or equipments that have to be lubricated should be designed so that they never, (never), do not leak oil or grease onto food or food contact surfaces. The equipment should be designed so that bearings or gears are positioned outside the equipment or sealed in a way that food is always protected. Only non-toxic food grade lubricants should be used on and for near food processing equipment.

Hygienic seals for food equipments.

Equipment should be designed so that the least possible supports are needed to meet safety and weight supporting requirements. Support legs should be designed so that they can be sealed and will not accumulate dirt, food debris or water where they meet the floor and the area can be easily cleaned and sanitized.

Electrical wires of the connection's equipments should be maintained together (as a flower bouquet) and where possible placed inside smooth conduit that will protect the wires from water during cleaning processes, so in that way, the area can be easily cleaned without producing a hazard. The electrical motors should be enclosed in a material that is durable, will not rust or flake, and can be easily cleaned. Also, the inside of equipment and pipes should have easy access so that they can be cleaned and scrubbed with a brush. Pipe joints may be needed to gain access for cleaning, and should have a sanitary metal pipe fitting with no internal threads.

The "difficult to clean nuts and bolts" (dcnb), of all equipments, should be in a minimum number if is not possible to design equipments without any "dcnb". These pieces accumulate dirt, food debris, and water and provide perfects microenvironments for pathogenic microorganisms to live, grow and contaminate your food. Nuts & bolts are forbidden in food contact surfaces, as a GMP requirement and as a logic thing to prevent. This threads increase with time of use and time of "life" of the equipment, so keep this in mind if you think to buy some used machinery, is better to do the effort of buying new and adequate equipments.

To meet these GMP requirements for the proper design and construction of equipment, utensils and food contact surfaces you need to check out:

What material is the equipment or utensil made of? Metal? What alloy? Plastic? What polymer?
Is this material the best choice for the task while balancing cost, ease of maintenance and durability requirements? Stainless steel better than PVC!
Is the equipment, utensil or food contact surface designed and constructed properly to prevent food from being contaminated when it is used? Turn on and see!
Can the equipment be easily taken apart and cleaned and sanitized? Try on and think!
Can you evaluate all the existing equipment to detect signs of corrosion, pitting, scarring, cracking or other deterioration that could harbor pathogenic bacteria and be difficult to clean and sanitize? Make a census not a survey, everyone of your equipments has to be listed, recorded, archived!
Could you develop procedures for fixing, removing or discarding equipment that is damaged and cannot be kept clean and sanitary? Write your own procedures, please!
Can you monitor all deliveries of new equipment to make sure that it meets your company GMP requirements? Take the custom to do so, as an ordinary thing in your own food business, with professionalism and ethics!
Can you monitor the condition of all your existing equipment, utensils or food contact surfaces periodically to make sure that they are suitable for use and will not contaminate food? Monitoring! Yes, your own autocontrol daily sheets!

Where to locate and install the equipments?

As a GMP requirement, all equipments should be installed and maintained in order to facilitate the cleaning of the equipment and of all adjacent spaces. <u>The equipments should be placed in locations where it can easily be cleaned and sanitized</u>. All equipment including ice makers and ice storage equipments should not be located in any area that is exposed to contamination from the outside environment, open or unprotected sewer lines, bathrooms, drainages of process water, and other contamination sources your food business local has.

It is of the most important to have adequate spaces between equipments in the processing rooms or lines, according to your layout and the disposed area dimensions. There must be <u>allowed enough breaches (major to 3 feet) in any aisles or spaces between equipment units or between work stations</u> or processing lines. These aisles must be kept open without clutter, garbage or other obstacles so that employees have room to do their work without contaminating the food they are handling and so that it is easy to complete a routine and effective cleaning and sanitizing daily goal.

About <u>the tabletop equipments</u>, these that are located on counters or tablet in your restaurant, grocery store, organic market or *gelaterie* must accomplish these features:
• Be <u>removable to facilitate the daily cleaning</u> and sanitizing. These equipment can <u>weigh a maximum of 30 pounds each</u> one, and <u>not bigger than 3 feet</u> (width, height or long), and without fixed utility connections.
• Be <u>sealed to the table or counter to prevent possible accumulation of water</u> and food debris under the base.
• Be <u>mounted on elevated symmetrical and stable legs for at least 4 inches above the table or counter to make easier the daily cleaning</u> +sanitizing.

If you own a little, medium or bigger food processing plant and works with automated manufacturing systems, the GMP program will requires from you (owner, manager, worker), that all holding, conveying and manufacturing systems, including gravimetric, pneumatic, closed, and digital/analogical automatic systems, shall be of a design and construction that enables them to be maintained in an appropriate sanitary condition, thinking in your entire processing system, cause your products move from one equipment or workstation to another using a system of conveyors (motorized, pressure driven, or gravity driven conveyors), and these also have to be maintained in the best possible cleanness and sanitary (disinfected) condition.

The conveyors and continuous systems of some food processing machinery and processing lines can be awfully intricate. It is important that all of the components in these systems that can have a possible contact with food meet the sanitary standards described earlier. Any areas where food residues, dirt, or water can collect can be a place where dangerous microorganisms can live and grow. For example, if a conveyor belt has hollow rollers, food residues and water will collect in the hollow space, making it difficult to reach during routine cleaning and sanitizing.

If bacteria grow in the water in this space, it could spread out when the rollers spin and contaminate the conveyor belt and any food that passes over it. Conveyors with only solid rollers should be used in a wet food processing environment.

The cleaning + sanitizing

The most important consideration in your food business should be given in how the system will be maintained and routinely cleaned and sanitized. Multipart systems may be difficult to break down so that all parts can be cleaned and sanitized individually. It may be necessary to drain cleaning and sanitizing solutions through some pieces of equipment while others may need to be disassembled. The system also needs to be located and installed in an appropriate place that will not interfere with cleaning and sanitizing, and with the favorable floor slope to avoid recontamination.

To accomplish this GMP requirement for the proper design, installation and maintenance of equipment, holding, conveying, and manufacturing systems your food business will need to develop <u>a checklist for evaluating all new processing lines or systems including conveyors and holding units</u> before they are purchased.

<u>Monitoring Cold Storage</u>
As a GMP requirement, each freezer, refrigerator and cold storage compartment used to store and hold any food that is capable of develops the growth of microorganisms <u>shall be fitted with a temperature control</u>, manually with a portable thermometer, (if digital, check battery charge daily, most errors comes from low battery gadgets) or temperature recording device if is the case, both of these controls are meant to show the temperature accurately within the compartment, and could be fitted with an automatic control for regulating temperature (thermostat) and you will have to set a certain level of temperature depending of what you store or maintain there to create alarms systems for each cold/freezer equipment: raw seafood, prepared "ready to fry" frozen food, frozen vegetables, dairy products, self-service food in costumer areas, etc.

The total control of your food business's rooms, caves, freezers, refrigerators, temperatures, daily, is of the most important and common way to minimize the growth of harmful bacteria and other undesirable microbes.

Food at proper cold temperatures prevents microbe's multiplication of the natural microflora all food have. <u>Cold (under 5°C) does not "kill" bacteria, viruses, molds or yeast, no, this physical parameter just slows their metabolism</u>, although some kinds of bacteria and molds could growth at temperature under Zero degrees Celsius (0°) like -18°C, and could recover in number if a small failure occurs in a cold apparatus of your food business. Food could deteriorate (you loss money) or worst, harm or kill people (you loss your company, and in certain countries your freedom). For this reason, it is <u>important to know and maintain the temperature of your different tasks coolers, refrigerators or freezers every day!</u>

The food tool here is the thermometer!

<u>Thermometers</u> to measure ambient air in your coolers, refrigerators and freezers <u>must include these features and you must carry out some crucial actions</u>: be smooth and easily cleanable; be durable for its intended, repeated use; indicating thermometers for refrigeration units must be accurate to ± 2°C; product thermometers used to test food must be accurate to ± 1°C; **must contain no mercury**; be accurately calibrated at least weekly to guarantee that they are operating correctly; be monitored by a designated and responsible person to check all refrigeration or freezer units daily to make sure that they

have appropriate temperature measuring devices that are working properly, and recording the checked temperatures in autocontrol sheets.

Portable digital thermometer for food business purposes.

Easy calibration procedures for digital thermometers
The very owner or worker of any food business can carry out these two simple methods to assure the accuracy of their thermometers:

i) with the "ice method", in which you will fill out a clean+sanitized glass recipient (25 cm height, cylindrical or cubical) of crushed ice, adding cold, clean water to fill the air spaces, then you will submerge the thermometer tip of the probe into the very geometric centre of the ice+water container, leaving for exactly 60 seconds. You should be reading at 0°C by now! In the contrary case, you must adjust it, leaving the probe submerged in the same position into the ice+water recipient until the reading be 0° C. Please take note of the date and the hour of this calibration in your calibration sheets. Take note of the room temperature and the outdoor temperature. Like this your digital thermometer will be "calibrated" each week, so your own controls on refrigerated, frozen and cold holdings units of your food business will be alright. Also, remember to check out the batteries of your gadget, a current source of errors in this matter..

ii) with the "boiling water method", in which you will heat a glass container (the same than before, be sure it is fireproof or heatproof) with clean water, using a constant source heat. When the water begins to boil you can insert the thermometer probe in the very centre of the water confined, but just 5 cm below the boiling water surface. Idem than before, please wait for 60 seconds. You should be reading 100°C! Knows that boiling water is affected by the atmospheric pressure (altitude), there are tables for standard readings of boiling water related to the altitude where your food business is.

Other food tools, the monitoring gadgets.
Apart from thermometers and cold temperature controls in storing and holding food units in your company, the GMP plan will require from you as the owner, responsible worker or supervisor, manager, <u>other instruments and controls for measuring, regulating, or recording heating unit's temperatures, pH, acidity, water activity, chlorine in water, or DNA</u> optical surface scanning gadgets looking for organic debris indicating bad cleaning +sanitizing!

The water content (and its real availability, free molecules of water), the carbohydrate, protein or lipid contents, the time of harvesting (vegetables), sacrifice (animal carcasses) or capture (fishes), the type and time of storing or conditioning, or the heat and/or cold treatments would modify the responses of the natural microbes all food have.

Certainly, as we cannot see actually the microbes growing, some parameters can help us to discover any deviation of microflora, changes of freshness of meat, fish or vegetables:

A) the parameter of pH will tell you, numerically, between 0 to 14, being 7 neutral, higher than 7 up to 14 alkaline, lesser that 7 down to 0 acid; as each food type has its own pH "fingerprint" (got to know this, for each one of your business products! record it in your sheets), any deviation can tell you if microbiological growth is happening (also odor, color test can help additionally), so get interest in acquiring a pH meter and take surveys on each product batch you receive, process and sell on this important parameter, taking records and calibrating it weekly with its own reference kits these gadgets bring with.

pH meter for food business purposes.

B) the parameter of the amount of available water (or water activity, aW) is more abstract than pH for non trained people, but believe us, if your food company works with glass bottled vegetables, fruits, meats and other food subjected to be preserved without cold storing, then you might acquire an aW gadget to measure the Water Activity and "see" if any microbial group can attack your processed food ready to distribute. Bad practices of preparation (less or more salts, additives, sugars), of cold storage (over freezing temperatures that get drought the surfaces of your meats, fishes, affecting the whole piece water balance), or the microbial growth itself can alter the aW parameter, allowing you to alarm the control systems you might have developed well, to avoid the processing or distribution of any "no conformity" food. There are normal aW reference levels for this parameter in each food, search the Internet and teach yourself in this important matter.

aW portable sonde.

C) the parameter of the amount of salt content in your food products, it is more associated with the aW mentioned before, in the way the salt ions acting as electrolytes could bond chemically and physically with water molecules, thus affecting the quantity of free water in the food of your interest, and free water content available means different rates of growth for microbes as bacteria and molds. The salt content at a processing step or in the final product must be controlled very well to prevent or minimize the growth of potentially harmful microorganisms in the food that is being produced. Remember that pH, water content and water activity, or salt level may be essential barriers to microbial growth in the finished product; the GMP requires that the instruments used to measure these conditions be accurate, adequately maintained, and appropriate for their intended use. You will need to follow the manufacturer's directions for their use, maintenance and calibration. Any instruments that are used should be designed for use with food products and certified for that.

Salt meter for food business purposes.

Good Quality Water supply

3° GMP's for Good Quality Water supply

Even in "dry" food industries and business with lesser use of "process water" like cereal mills or sugar factories, for example, there is always the need to rinse (water), clean (water+detergent), sanitize (water+sanitizer) and rinse again (water) the equipments, surfaces, tools, rooms and its floors and walls affected by the food processing. So, an excellent and reliable and constant source of a good quality water supply is crucial to your food business.

Although, any water is free from ubiquitous microbes (even in hot water springs or gelid Antarctic waters, exist vast numbers), your company could achieve reserve tanks of good quality process water and human consumption water in the facilities where you will need this important element. Be sure of putting various obstacles to microbe's growth like chlorine at accepted levels, ozone treatment, ultra filtration stones, and/or UV radiation in your water reservoirs. Try to not rely in a sole obstacle because microbes tend to acquire resistance to it: *an example, we explain graphically in our courses, is a boxer (food business owner) with several types of punches that could defeat anyone (microbes), but a boxer with a single and unique jab punch, will discover that its opponent gets used to it, and develop strength each time higher, thus creating a big problem.*

Have a layout of your water pipes, sources and reservoirs network of your food business; guard it very well and make regular (monthly, weekly?) visual surveys to observe: possible leaks, oxide spots in pipes, backflow, external biofilm (microbes) colored spots, or bad functioning of water counters and taps.

You must also follow, mandatory or not in your country legislation, the recommendation we tell to our clients: to carry regular microbiological samplings of water used for human consumption and for process water (even for laboratory water if you have Quality control labs using distilled/deionized water) with third party companies that execute a sampling planning for you and with you, in coordinated services fulfilling the quality features of a scientific sampling for microbiology: disinfected (sterile) bottles, a source of heat (flame), sterile gloves for sampling personnel as well as hairnet caps, mask, rubber boots and the proper white coat. You might profit to carry some lectures of pH, temperature, chlorine levels in the process water you will sample, to make correlations with microbial diversity and its charges.

If your food business is small, make a total water points sampling (restrooms, handwashers, dishwashing facilities, hoses, reservoirs, and even icemakers); if you own medium to large food businesses, you might do proportional water points sampling or with the minimum number your country's legislation states, 5 water points each month as in some countries.

We consider two types of process water in any food business: direct and indirect process water. Lets see: direct process water is the used in defrosting some frozen pieces of fish, meat or poultry before, that is "touching" your food product, as well as rinse water when you clean and sanitize the surfaces of equipment, conveyors, tools and any device with any contact moment with your food. Crushed ice beds to store prawns, shrimps, fishes, and other food items in cold caves until processing (a typical action to avoid cold dehydratation in your food), is seen as process water, and you have to control the microbial charges in it or in the water which is made of, be in the same company facility or if you have an ice supplier (in this case you might demand that supplier for the respective microbiological controls of its ice products). Indirect process water is, for example, the water of hand washers everywhere in a big food plant or in the restrooms at any restaurant, which will help in cleaning and sanitizing our hands every time our professionalism dictate us: so, a bad quality "tap water" in a hand washer point, would be a risk to the safety of raw food you are manipulating in the line of production or in the desk of your restaurant's kitchen, because your hands would be possibly contaminated with microbiological charges, and then you touch your food products, so, indirectly, the tap water is "touching" your precious food.

You can find in the internet some guidelines for water supply requirements directed to Good Clinical Practices or Water for Pharmaceutical Use, even for Aquaculture there is need to control water and seawater quality systematically, well, as we have mentioned before, the task of food handlers, food business owners and workers, is very similar of that of surgeons or pharmaceuticals, they will demand the best source of excellent quality water to their process (to wash hands of surgeons before a caesarean section, or a liquid medicine using water as ingredient, to clean surgery kits or rinse the equipments of medicines factories).

Not only microbes are dangerous hazards to your food business's process water, but dozens of chemical substances and elements that are forbidden, totally or permitted up to a limited value: your process water must not have detergents, soaps, sanitizer, pesticides, excess of chlorine or any salt, harmful heavy metals, synthetic oils, greases, lubricants, nitrogen or phosphorus contents, etc., above the permitted by the local law (research what your country demands in these subjects, please, these rules tends to change in each corner of the planet). Physical features as odor, color, taste, turbidity, and suspended solids, etc., would reveal major problems of chemical and microbiological origins. These are easy to control with your senses (eye, nose, *goût*) and with simple digital devices as turbidimeters.

Returning to microbes, but in human consumption water (a number of drinking water devices you must have accordingly to the staff size of your food business). We present a United States federal drinking water regulation for microbes, where there is "Zero Tolerance" to coliform bacteria, viruses and *Giardia lamblia* (a parasitic protist); zero tolerance means that the Maximum Contamination Level (MCL) is zero (0 coliform bacteria per sample of drinking water) just permitting a 5% of monthly samples presenting this indicator, and, in those cases, it must be investigated the same sample or point of process water at the food business, for fecal coliforms and for *Escherichia coli*, one of the most famous fecal bacteria in the planet (in vertebrates animals and "invertebrates with guts", think about it!), but not the unique. For example, other regulations around the World includes different fecal bacteria in drinking and contact (with human) water, like *Streptococcus faecalis*, a "more human" faeces related bacterial species, with other congeners species more related to caw (*Streptococcus bovis*), horses (*Streptococcus equinus*), dogs (*Streptococcus canis*) and go on other ones. What we want to attire your attention here is, that there are different regulations cause there are different realities in any country around the World, based in its degree of public sanitation, its latitude and climate, the biodiversity of the country, so, if in the US regulation ask you for coliforms, viruses and one single parasite, in India there will we slightly different microbiological parameters, as well as in Venezuela, Nigeria or Turkey.

You must assure to your costumer wealth, the most safe of drinking water that your may serve for free in some cases if you owns a restaurant and give a glass of water to a child sitting in family at your disposed tables: kids, as ill persons, pregnant women and elder people are categorized as immune compromised as they does not have the strength of an adult, healthy enough to fight (its body cells and tissues) against possible infections from any microbes that conquer its gastroenteric system and infect (r worst poison the person). If you do not want to harm anybody, follow the GMP recommendations.

Our founder, Dr. Luis Elguezabal Aristizabal, used to promote the "combined hurdle method" to control efficiently the microbes in ingredients, raw food, fruits, vegetables in the food companies he advised, and this smart method will be very effective as well in controlling the microbial dangers in our water reservoir tanks that supply your process and consumption water in your food business.

This is, to combine many sucessive or simultaneous hurdles or obstacles to microbial growth and development, and not to rely in the most famous (cheap?) control available, chlorine.

Yes, chlorine is helpful, but at prolonged use it can be cause of microbial resistance in the inner surface of your tanks, water pipes or hoses, and creates a big problem: the biofilms. A touchable and visible accumulation of trillions (in each square centimeter!) of bacteria, algae and/or fungi that will pose an eternal source or focus/niche of contamination to your process water, and then your food products, food business facilities you "clean" daily with that water. Then, the food business owner tends to increase this control, "more chlorine"!, and the natural selection still will be acting in behalf of the microbes, making them stronger each month against this "control" they are getting used to it.

What to do? well, in urgent actions you can purge and clean in place (CIP) your whole system of water pipes, tanks and hoses, with special soaps and sanitizers (even you might use organic acids like citric or acetic acid), or, you might prevent the "difficult to clean" biofilms by using diverse controls in your water reservoir tanks, like: i) UV radiation bulbs, secured in the top of the tank inner "roof", combined with ii) ozone treatments, iii) previous ultra filtration of water entering the tank, and iv) of course the cheap chlorine. You might combine and adjust the levels of each one of these hurdles, in a way that could prevents the resistance developed by microbes in your food business water.

- Personnel
- (Hygiene Controls
- /Continued
- Education)

4° GMP's for Personnel (Hygiene Controls/Continued Education)

The "human resource" is the most important you have and will have always in your food business. This is not an exaggerated sentence, no, it is our (the founders of A6 and us) own experience and the common pattern/trend observed in any of the food business clients which we have advised, trained or carried out microbiological samplings: canned seafood megafactories, frozen seafood companies, dairy plants, maize and wheat mills, restaurants, supermarkets, bakeries, meat processors, and other kind of food related business.

The persons working as food operators, as food operation supervisors, as food business managers, they all, with direct and indirect contact with raw and/or prepared food products are considered the "Personnel" in your food business, without any distinctions, cause we all have skin, mouth, nose, ears, eyes, genitals, and we all go to the bathroom, and we all are potential dangers to our food products if we act with riskiness actions all the time!

If we have bad hygienical habits or do not wash our hands properly every time we go to piss or the other "thing", be at home, at work, or in a restaurant when you invite your couple to dinner, you are accumulating probabilities of increase your "bad" microbes numbers in your hands, skin, clothes, etc. Hands are our very precious tools in food processing business, and to do that simple but important thing of good hygiene for ourselves as a natural thing is crucial, also, if we assume this act as responsible/professional reflect, as we were surgeons as we stayed before, but with the different that a sole medical doctor (or team) is responsible for one and only one patient at a time, and we, the "food handlers", we are responsible of dozens, hundreds or thousands people that trust our food products, companies and brands, as safety, innocuous, healthy food supplier for all persons including child, pregnant women, ill or elder people. In the final chapter of this document we will inform you about the food hygiene and food microbes that could harm us and harm our clients.

Follow this *Decalogue* or checklist for food business's personnel hygiene:

Disease Control
No sick workers rule
Prevention of contamination (open sores, wounds)
Good hygienic Practices
Hand washing and personal cleanliness routine
Hair restraints, no jewellery or loose articles regulation
Mandatory "No eating, drinking, tobacco products in production"
Recommended "Proper attire" (gloves, aprons, uniforms)
Continued food/personal hygiene trainings
Regular microbiological skin/hands surveys sampling for cleanness verification

Disease Control: Personnel known, or suspected, to be affected from, or to be a carrier of a disease/illness likely to be transmitted through food (leaking nose, fever, skin eruptions, cough, "red eyes", etc.), should not be allowed to enter any food handling area if there is a minimal chance of get food contaminated. Any person affected in these ways must (him/herself!) immediately report illness or symptoms of illness to the managers and supervisors. Medical examination of a food handler should be carried out if clinically or epidemiologically indicated, and to assure its proper comeback to work.

This person could not work today and until he/she recovers a healthy appearance.

No sick workers: jaundice, diarrhea, vomiting, fever, sore throat with fever, visibly infected skin lesions (boils, cuts, etc.), discharges from the ear, eye or nose, and other possible health affections of any personnel, poses the responsibility of managers and supervisors that should whom must be reported from the affected worker or its colleagues, to send for medical examination and/or immediate exclusion from food handling until this hygienic risk is controlled.

This kind of skin lesion is inadmissible in a food handler or food worker of any area (supervisors, managers, and administrative personnel).

Prevent contamination (open sores, wounds): although our body produces anticontamination reactions, "sending" millions of blood white cells to each wound we have in any part of our skin, killing, engulfing the bacteria and fungi that could infect our body, these wounds represents a concentration of millions of bacteria or fungi that are "eating" our outer body tissues, and if this microbes develops in the proteins and lipids of our tissues, would be develop as well in any food product they could reach by any chance of random actions we take, by being wounded at (a food business) work! Do not come to work if you have open wounds in your hands, arms, and head.

A large sore, altough could be controlled and sealed, it is better to recover the proper skin health before you or your food personnel come backs to work.

Hygienic Practices: food personnel with unclean hands, and exposed portions of arms or fingernails, sweat from face and hair, secretions, can contaminate your food. If your consumer clients eat contaminated food, foodborne related illness or outbreaks may result. One of the most important actions for food personnel is to wash their hands, and should do this immediately after participating in some activities that certainly do contaminate the hands. The washing of hands should be done when: i) entering a food preparation area, ii) before putting on clean, disposable gloves for working with food products and between each glove changes, iii) before participating in food preparation,

iv) before handling clean equipments, tools, cutlery, silverware or food serving utensils, v) changing tasks and skipping between handling raw foods and working with "ready to eat" foods, vi) after handling soiled dishes, equipment, or utensils, vii) after touching bare human body parts, ears, hair, nose, face or exposed portions of arms, viii) after using the bathroom in any of "our two physiological needs"; ix) after coughing, sneezing, blowing the nose, using tobacco, eating, or drinking, x) after touching any domestic mammal or aquatic animals such as shellfish in display tanks, and even plants.

Food handlers should maintain a high degree of personal cleanliness and, where appropriate, wear suitable protective clothing, head covering, and footwear. Cuts and wounds, where personnel are permitted to continue working, should be covered by suitable waterproof dressings. Personnel should always wash their hands when personal cleanliness may affect food safety, for example: from starting of food handling activities; immediately after using the toilet; and after handling raw food or any contaminated material (floor, walls, pallets), where this could result in contamination of other food items; they should avoid handling ready-to-eat food, where appropriate.

- Washing you hands is too important. Not doing it well, is lethal.
-
- **Hand washing and personal cleanliness**: the handwashing steps to food personnel are resumed as follow: i) clean your hands and exposed portions of arms, including surrogate prosthetic devices for hands and arms, for at least 20 seconds, ii) rinse under sanitized, clear running water, iii) apply soap and rub over all surfaces of your hands and fingers together vigorously with friction for at least 20 seconds, giving particular attention to the area under the fingernails, between the fingers/fingertips, and surfaces of the hands, arms, and takeoff prosthetic devices, iv) rinse again thoroughly with clear and sanitized water, v) dry your hands and exposed portions of arms with disposable paper toweling, or a heated-air / high-pressure hand-drying device, vi) always avoid recontamination of your hands and arms by using a clean barrier, such as a paper towel, or available sterile gloves in bathrooms, when you turn off (and would touch, as many different people before!) hand sink faucets or touching the handle of a restrooms doors in your food business.

Hand washing facilities. Better invest in "foot-activated" hand washers.

-
-
- Additionally, all personnel must have a Doctor's certificate at the time of hiring, and renewed each year at least. Food businesses (maybe yourself) workers must inform their supervisor (or partner if it is your business: friend, family), that you are ill with some symptoms that could contaminate ingredients or products. Remember that no

medication is allowed in the food handling facilities, and ensure that clean bandages cover any open wounds you might have.

•

Hair restraints, no jewellery or loose articles: like all titles referred to personnel in food business, here is a must be "transversal line" involving all kind of personnel that a food plant or restaurant should have: the top-down approach. This ensures that personal hygiene policies and procedures are implemented by all personnel, every one, the managers, visitors, production workers, sanitation workers and maintenance staff—at all and each company's facilities. This worldwide "law" will reduce the risk of product contamination by foreing material and the likelihood that such product will reach the consumer. Remember, even if there is not a high danger when you find a earing in a home made cake, and prevent a possible asphyxia in your family, to found* (on time, before you eat it!) a ring in a chicken nugget, a hair in a baguette bread or a toothpick in a pizza slice is very dangerous for your....brand or company! That would decrease the respect and confidence the clients have developed with your food business. If your clients do not go any more to your pizzeria or kebab restaurant for these kinds of reasons, you better change of business. Food business of your own are good to work and gain familiar and individual liberty, but you must be very professional and rigorous, always. So use hair restraints, you and the personnel you direct in your food business, train this "human ressource" to not wear any other physical object than the appropiate clothing to food handling, and not to carry loose articles like pencils, cellular phones, coins, etc.

*all these examples, are some of my own experiences in the matter, while going out with my 3 kids and wife to dinner. The hair founding is terrible, it is also a microbiological hazard of trillions cells of bacteria, yeast, molds.... we never go back to those places of course.

Good example of a hair restraint. Not ponytails or hair bands.

No eating, drinking, tobacco products in production: sloppy feeding particles, drops of drinks and saliva, carries out million of different microbes species, some harmful. The smoke of tobacco you see and smell could easily impregnate the raw food of ready to eat food in your food business. These actions: eating, drinking, smoking, are very controlled in food businesses for food workers, owners, not only in the working hours, but at any hour and day your business is active producing, processing, transporting, handling, or vending good food!

Not only careless employee practices can cause product contamination. Any behavior that could result in food contamination such as drinking beer, eating donuts, use of tobacco, chewing gum and every unhygienic practice, is not allowed in food handling areas. The way to avoid these routes of contamination is to interdict it!

Remember that your clean clothes are important keys to maintain all your actions safely to food handling: the proper attire (gloves, aprons, uniforms), wearing pants and covered sleeves, with special shoes (neither open toes nor high heels) to be worn in your food business. The personal belongings and street clothing must be stored in appropriate locker rooms.

Not even a Chef should eat at his/her labour time (cooking).

Continued food/personal hygiene trainings: People engaged in food handling businesses should refrain from behaviour which could result in contamination of food, that is it: i) to commit ourselves in observe the logic of all contents explained in this kind of documents you are reading right now, ii) to follow all rules and laws in the food safety organisms (national + international), iii) to act professionally, morally, ethically every moment when handling food at any step of the chain at the company you work today, or the business you will manage tomorrow.

Continued training, formation, education, and pedagogical methods to learn all contents and information related to food sciences, updated and basic ones too, must be a mandatory rule in each food business around the planet Earth!

Remember that: smoking; spitting; chewing or eating; sneezing or coughing over unprotected food is unsafe (stupid by now), that your personal effects such as jewellery, watches, pins or other items should not be worn or brought into food handling areas as they pose threats to the safety and suitability of food you will sell or consume. Take baths daily, using no perfume, aftershave, nor fragrant creams. With the internalisation of each food worker/owner to not wear any jewellery, any false nails or nail polish, that your fingernails should be kept short and clean. Do not leave gloves, masks, etc., lying around while on break or at shift end.

See if crates, boxes, containers or buckets are not being placed directly on the floor, and if your food business accomplish with the storing of brooms and dust pans at stations provided. Report it if this simple principle is not followed.

An A6 open course (2011) of Good Manufacture Practices in Food Handling.

Help the supervisors of your food business and i) keep hands contact with ingredients to a minimum, ii) check ingredients for expiration dates to ensure that fresh ingredients are used all the time, iii) observe if cooling products are always be kept covered, iv) prevent to lean, sit or step on product surfaces, v) avoid to handle ingredients or products with either cut or infected hands, vi) keep contact surfaces clean and free of contamination from tools, cords, cleaning utensils, machine parts, lubricants and paper, vii) clean all spills promptly, viii) keep everything off the floor and the area clean and floors swept, ix) help in maintaining work areas that should be cleaned regularly throughout the shift, x) keep your immediate working area swept or dust mopped and wipe or mop up spilled liquids promptly, xi) scrape the floor around the work area after completing a task, xii) leave your work area clean at the end of your shift, xiii) ensure that all pallets and materials are kept at least 45 cm away from the walls, xiv) inspect torn bags and boxes and repair it if appropriate, xv) store ingredients and products at the appropriate temperature, and, very important, xvi) do not engage in funny playtime or joking, only in your breaktime hours, please. Food businesses are serious affaires.

Regular microbiological skin/hands surveys sampling for cleanness verification:

As part of auto control methodologies in many food plants following the GMP, HACCP and ISO food quality programs, routinary microbiological surveys are made in hands of direct food handlers (operators) involved in any processes and products of the food plant, as well as the cookers and Chef's hands of its "industrial" canteens who prepare meals all day for the food plant workers! This action is meant to the verification of the efficiency on hand washing among the personnel of any food business.

The reports are presented without names but with codes that supervisors would bond to every surveyed worker in any "hands sampling" of the week.

In our experience, this could lead to a better and personalized management of personal hygiene among the personnel of the food business. We use prepared Petri dishes with sterile culture media (total count of aerobic bacteria; molds and yeast; *E. coli, Salmonella, S. aureus*, or whichever microbial type you want to control) Also we have carried out sampling of surfaces on equipments, tools, cooking apparels, using sterile hyssops, and sampling of air during 30 minutes by opening the Petri dishes in a particular site of interest for the owner or manager of the food business: kitchen, bar, entrance, restrooms, etc.

Sterile hyssop for surface microbiological surveys

Air quality sample after 30 minutes exposure of a open Petri dish. Total aerobic bacteria (incubated 24h/37°C) Plate Count. Sugarcane mill, Cariaco, Venezuela, 2005.

Food Control
Volume 18, Issue 4, May 2007, Pages 326–332

The occurrence of indicator bacteria on hands and aprons of food handlers in the delicatessen sections of a retail group

J.F.R. Lues, I. Van Tonder

School of Agriculture and Environmental Sciences, Faculty of Health and Environmental Sciences, Central University of Technology, Free State, Private Bag X20539, Bloemfontein 9300, South Africa

A research article on this matter. It is real, it happens everyday everywhere...

> • Hygienical Requisites for Food Production (*SOP's)

5° GMP's for Hygienical Requisites for Food Production (*SOP's)

Although the term Standard Operating Procedure, or SOP, is used in a variety of different businesses, from medical, engineering, education, heavy industries or food industries, plants or small and medium businesses, we will refer to the later ones: Food SOP's.

FAO refers this key step in quality programs in its view as an important aspect. <u>SOP's are documents</u> well assembled and easy <u>to read and follow the instructions</u>, permitting that any personnel, food worker o new employee in the food business could understand all operations described in <u>the right order to produce</u>, not only safe food but <u>uniform and stable food quality</u> during a whole batch or lot of food product, during an entire time period and during the whole history of that product. For example<u>, think in a recipe</u> for a patisserie, the more "alchemist" of the food cooking worlds; a recipe is a group of ordered ingredients and instructions directed to obtain always the same product appearance, taste, texture and smell (and success), and if you fail one day, your clients could say "uhmmm this *Gateau Basque* .... is not the same... it lacks... vanilla.."

So, the purpose of a SOP document is to follow the operations correctly as you and your team have written down in paper, and always achieving the "same quality" in the same manner; these SOP documents must be available at the places where the work are done: one for each product you have, be "ark clams with hot spice sauce", be "Hawaiian salad tuna" or "sardines in soy oil", these are 3 SOP's documents in your canned fish food plant.

A SOP is an obligatory instruction, and if any deviations from the expected conditions have occurred, these should be documented including the person name that could give permission for commercializes this product, if possible. The original should rest at a secure place while working copies should be authenticated with stamps and/or signatures of authorized persons.

FAO share an example of <u>SOP's for a quality laboratory</u> in a food plant

-. Fundamental SOP's. These give instructions how to make SOPs of the other categories.
- Methodic SOPs. These describe a complete testing system or method of investigation.
- SOPs for safety precautions.
- Standard procedures for operating instruments, apparatus and other equipment.
- SOPs for analytical methods.
- SOPs for the preparation of reagents.
- SOPs for receiving and registration of samples.
- SOPs for Quality Assurance.
- SOPs for archiving and how to deal with complaints.

SOP's comprises detailed work instructions for a particular food product in your business while the basic GMP's (Food Good Manufacturing Practices) is a list of: i) actions to "do" and actions to "don't", focusing in riskies behaviors of food business personnel, ii) knowledges of all kind of dangers existing from chemical, physical and biological sources, iii) directives of best designs for food business installations and facilities, iv) instructions to manage all food products of your business with the same standard of quality referring to the food safety maximal standard, offering an accurate innocuity, harmless to the clients, consumers, the public.

An example of SOP. Dairy product: Fresh Cottage Cheese

I) You must ask your milk supplier a proper system to check incoming raw milk, assure that it is following an effective food safety programs, as GAP (Good Agricultural Practices) or an organic/ecological/biological certification production; also you might require this supplier for proofs of Hygienical Food Handling courses in the farm/dairy plant staff.

II) Develop, consciously, the written specifications for receiving appropriately the milk, the sampling at the arriving to check freshness (acidity; microbial total counts with rapid -minutes- microscope Breed smear method; or "instant" -11 hours- kits to detect lethal bacteria like the *E. coli* O157 strain).

III) Follow your developed SOP in the treatments to homogenize milk, skim at your preferred fat content, and pasteurize with the right times and temperatures needed (until reaching 72°C for 30 seconds, then turn off heat source). Let the milk to temper around 36°C. (By now you know you must buy a good thermometer!).

IV) As you or any personnel in your dairy business can read in your prepared SOP for "Fresh Cheese", the next step will be to add the rennet (enzymes from calves' stomachs, standardized, safest and best option) that promote the coagulation of milk solids (the main proteins, casein and lactoglobulines, that engulf also the lipids and sugars, minerals, vitamins of milk). In few minutes a beautiful layer of curd appears floating over whey*. You will cut the curd with clean knife like instruments if you don't have a lira. Pick up the cubes of curd still floating in the whey and pour them in a clean recipient for adding salts at your preferred taste.

V) Continuing reading your own SOP, once you have salted the mass of curd, it must be placed it at the mould or press, a clean and sanitized cylindrical stainless steel, perforated (to drain excess water/whey) and with a pressure device or "dead weight" above to allow downsizing at preferred level of water content.

VI) Now your know that the next step is to package the fresh cheese just made, in proper packaging material, always clean and disinfected, as your hands and instruments during the whole process of making your cheese. * With the remaining whey you have, you might want to make Ricotta cheese: develop another SOP for this!

An artisan delicatessen, fresh cheese from Carora breed caws. Produced with SOP's guidance.

So, for SOP's you must think in all ingredients, food grade packaging, chemicals, tools, instruments, equipments and all the necessary to obtain any projected product in your business. Including the inspection procedures at any steps or critical points you must control (read HACCP chapter examples) and assuring the written procedures for handling and storage of raw material, ingredients, additives and final product are well transmitted in these important documents.

SOP's are an easy way to perform the described Process Control, that you or the official authorities demands. If food business personnel reads that a proper temperature, clean conditions, and sanitization procedures must be controlled, everything will be in the good rail. Also write the procedure and aim of the First In- First Out (FIFO) logic, and explain the reasons why to do this FIFO thing on storing facilities, be of raw material, additives, ingredient or final product again, all of which must have a recording sheet system to write down the dates and log incoming products to the storage units.

6º GMP's for Sanitization Programs (*SSOP's)

SSOP's are documented steps that must be followed to ensure adequate cleaning of product contact (and non-product) surfaces. The cleaning procedures must be very well detailed to assure that any adulteration of product will not occur. Every HACCP plans require a SSOP's guide to be documented and reviewed periodically to incorporate any changes in the physical structure of the food plant, and carried out by responsible well trained worker/supervisor/manager. The SSOP's procedures can be publicized openly and may be accessed by the public or an official institution, for verifications of the records and its possible serious failures in it.

SSOP's + a Master Sanitation Schedule + a Pre-Operational Inspection Program = Integrated Sanitation Operational Guidelines for food related processing business and one of the primary backbones of all food industry HACCP plans.

SSOP's are very simple but sometimes also very difficult. Food industry equipment should be constructed from de beginnings with a sanitary design. Each equipment and area must "have" an individual SSOP and this should describes:

- The equipment or affected area to be cleaned, identified by common name,
- The tools necessary to prepare the equipment or area to be cleaned
- How to disassemble the area or equipment
- The method of cleaning and sanitizing

Responsible persons

Who has the top responsibility? You. As a food business owner/manager/worker. You are responsible for activities related to implementing and maintaining the SSOP's documents and activities described there in for each area, equipment or tool as your Sanitation Scheudle (which you have constructed) dictates.

What must be done to implement and maintain the SSOP? Implementing and maintaining the SSOP requires:
- you, your team or your supervisor must review the SSOP's documents regularly,
- doing the daily monitoring of pre-operational and operational SSOP procedures,
- recording the findings of monitoring,
- performing or assigning any corrective actions necessary, and
- documenting the corrective actions.

Easy!

Can SSOP responsibilities be delegated? Yes. Only the food business owner or manager or sanitation supervisor could assign this crucial responsibility of SSOP's obligations to other trained personnel. Record names and dates of any changes in the responsible for SSOP's worker.

<u>When should the SSOP be signed and dated?</u> All SSOP's must be signed and dated when executed and when it is modified.

<u>Who should sign and date the SSOP?</u> The food business owner or manager or supervisor of Sanitation must sign and date the SSOP, depending on the size of your business, maybe yourself should sign, supervise, and execute!, the sanitation procedures at the right moments (for example, when the food plant stops production or your restaurant is closed to the public).

<u>How much time must SSOP records be archived?</u> All records of to the SSOP's will be kept in offices archives at the plant for a minimum 48 hours. After those 48 hours, the records shall be kept for at least 6 months at another archive in the same building.

<u>What must be done if Official (government, sanitary office) personnel ask for the SSOP records?</u> All SSOP records will be made available to Official personnel (within 24 h) upon request.

Pre-Operational Sanitation

<u>What is the general procedure for sanitation of food-contact surfaces?</u> All equipment and all surfaces that could be in contact with food or additives, ingredients must be <u>cleaned and sanitized at the end of the shift in which it was used</u>.

1. Disassemble the equipment. Place the parts in the designated tubs, racks, etc. as well as all simple equipment and hand tools that are cleaned and sanitized in the same manner, without any disassembly and reassembly.
2. Physically remove product debris by hand or with tools such as scrapers.
3. Observe equipment for missing parts or parts/surfaces that are worn to the extent that debris will accumulate and cause product contamination. Replace or repair parts/surfaces and document what was done in the Corrective Action Sheet.
4. Rinse equipment parts with the <u>cleanest potable water available</u> to remove remaining debris. The source of municipal water, should be tested on national laws refferring potability of water to food contact and/or human contact, we propose, monthly, to check out coliform bacteria, molds, yeasts, excess chlorine, odor, taste, color, pH, etc.
5. Apply an <u>adecquate soap + sanitizer</u> to parts and clean according to manufacturers' directions. It is (HACCP) recommended to clean floors first and then clean equipment from top to bottom and in the reverse sense of layout flow, from cleanest (more critical, final product) to dirtiest parts of process at the begining of process.
6. Rinse the equipment parts with best possible quality potable water.
7. Sanitize equipment with an approved sanitizer that is mixed and used according to the manufacturers' directions, and, always, rinse with potable water again (<u>traces of soap or sanitizer are seen as "chemical dangers" to your product quality</u>!).
8. Check and reassemble the equipment. Note that some equipment surfaces will

be sprayed with white oil (to prevent rusting) before reassembly.
9. All cleaning and sanitizing chemicals shall be properly labeled and stored separately from food and processing areas.

Small food businesses clean + sanitize "equipment".

How do we monitor the sanitation of food-contact surfaces?

The Plant Manager will inspect equipment and other food-contact surfaces before the start of production each workday to monitor the effectiveness of cleaning and sanitizing. The Plant Manager will normally rely on appearance, odor, and feel of food contact surfaces (an "organoleptic inspection"). Any necessary corrective actions should be performed and documented in the Corrective Action Sheet. The corrective actions taken must prevent direct product contamination or adulteration. If new inspection procedures are adopted, the SSOP will be modified accordingly, signed, and dated.

What is the general procedure for sanitation of surfaces that might have indirect contact with our products?. Even if the SSOP guidelines do not explicitly address potential indirect food-contact surfaces such as floors, walls, and ceilings, these surfaces can be an important source of microbial contaminants. We regularly perform the following steps to maintain sanitary conditions.

   a. Sweep up debris and discard it.
   b. Rinse surfaces with cleanest potable water available.
   c. Clean surfaces with tested soaps and sanitizers, according to each manufacturer's directions. Dilutes in the same water quality used in rinsing (process water).
   d. Rinse surfaces with potable water.
2. Cleaning Frequency: Clean processing area floors and walls at the end of each production day or if it is very dirt before that, perform it before each time wolud be required. Clean ceilings at least once a week, and more often if needed. You must plan the cleaning frequency and have it writen down in the specific SSOP's chapter.
3. Also clean and sanitize the cooler/freezer floors, walls, and ceilings. Shield or remove product before cleaning to prevent it from being splashed. Follow the Cleaning Procedures described before.
4. If cooler/freezer shelves and racks are in need of cleaning and sanitizing, remove product and clean using the Cleaning Procedures described before.
5. The Pest control is done by a third party or commercial applicator. The commercial applicator will provide a record of his/her inspections, findings,

and actions taken. These records will be kept on file in your company archives. The plant manager will monitor plant entryways on a daily basis during production to assure that insects and rodents cannot enter the plant. Rodent traps will be monitored daily to ensure that they are properly placed. All pest control chemicals shall be properly labeled and stored separately from food/processing areas.

How do we monitor the sanitation of potential indirect food-contact surfaces?

The Plant Manager will inspect potential indirect food-contact surfaces before the start of production each workday. The Plant Manager will normally rely on appearance, odor, and feel of indirect food contact surfaces (an "organoleptic inspection"). Results of the inspection will be recorded on the SSOP Inspection Form. Any necessary corrective actions should be performed and documented in the Corrective Action Log. The corrective actions taken must prevent direct product contamination or adulteration. If new inspection procedures are adopted, the SSOP will be modified accordingly, signed, and dated.

How do we use the SSOP Inspection Sheet?

Record the inspection results on the SSOP Inspection sheet. If an inspected area, program, or piece of equipment is acceptable, enter the appropriate symbol (✓). If a deviation is noted, enter the (X) symbol in the SSOP Inspection sheet, and then describe well the problem and the corrective actions taken to fix it on the Corrective Action sheet. Be sure to date and initial these records. The corrective action may consist of re-training the sanitation crew employees as appropriate, changing a cleaning/sanitizing procedure, and/or repeating the existing procedure with greater care and re-inspecting.

Operational Sanitation
The objective of our operational sanitation program is to prevent contamination of raw food, food ingredients/additives and other food products resulting from employee actions throughout processing.

What sanitary practices must be followed by all employees?
1. No person with illness, or open/infected wounds is allowed to handle foods or food-contact surfaces.
2. All employees must begin their shift wearing clean garments. Raw product processing employees must wear hair covers and change or clean/sanitize (or replace) outer garments when they touch soil, walls or any other inapropiate surface. Ready-To-Eat (RTE) product processing employees must wear hair covers and single-use disposable gloves, and maintain the cleanliness of all outer garments.
3. Employees must **wash hands properly** after using the bathroom or handling any objects that may contaminate products, and before putting on disposable gloves.
4. Employees must not use tobacco, eat, or drink in production areas.
5. Employees may not wear any jewelry or cosmetic items that could contaminate product.
6. Food, beverages, and medications must be stored in designated employee locker or storage areas.
7. Hand wash facilities and toilets must be kept functioning correctly and properly

<u>supplied. Also, as clean and disinfected as production areas!</u>

SSOP Sheet: Previous and during SSOP's

Pre-Operational items (to be monitored each day before *turn on* the food plant)	Daily Control Enter: Y= conformity, N = Unconformity Name of respnsible person besides each item.
	Date:
Food processing area and product-contact surfaces must be in good order, and are cleaned and sanitized after operations.	
Processing area product-contact surfaces are in good order, and are cleaned and sanitized after operations.	
Equipment and facilities that are potentially indirect food-contact surfaces are clean and in good operating condition.	
All cleaners, sanitizers, pesticides and other potentially toxic chemicals are properly labeled and stored separately from food and processing areas.	
Rodent traps are loaded and properly located; entryways prevent rodent entrance.	
Food containers, packaging and dry storage areas are maintained to prevent direct or indirect contamination of food.	
All food transport equipment is clean and in good repair.	

Operational items (to be controlled and recorded at each shift of any given work day)	Daily Control Enter: Y= conformity, N = Unconformity Name of respnsible person besides each item.
	Date:
No person with illness, or open/infected wounds is allowed to handle foods or food-contact surfaces.	
Employees do not wear jewelry (other than secured wedding bands) or cosmetic items that could contaminate product.	
Employees are wearing clean garments, gloves and hair covers (as necessary for assigned tasks).	
Food, beverages, and medications are stored in designated employee locker or storage areas.	
Employees wash hands properly after using bathroom or handling objects that may contaminate products.	
Employees do not use tobacco, eat, or drink in slaughter or production areas.	
Hand wash facilities and toilets are in good supply and functioning correctly	
Tools, hands, aprons, and boots are cleaned and sanitized (if appropriate) to prevent contamination during evisceration or during processing of skinned carcasses.	
Brisket saw is rinsed and sanitized before next use.	
Tools that have potentially contacted SRMs are cleaned and sanitized before next use.	
Appropriate scheduling, separation, and/or cleaning/sanitizing procedures are used to prevent cross-contamination with allergens.	
Dry and wet waste materials are properly contained and removed from the processing area. No accumulation of waste materials.	
Ready-To-Eat foods are handled and stored such that they are separated or segregated from raw foods, raw food containers and packaging.	
Work surfaces are cleaned and sanitized between handling different foods, or between raw and Ready-To-Eat foods.	
Condensation is removed from process areas in a sanitary manner.	
Other:	

SSOP's – Sanitation Standard
Operating Procedures
• SSOP's guidelines should be in place for the cleaning and sanitizing of all food processing equipment.
– Need to register everything you and your personnel achieves everyday on this matter.
• A Core Central Sanitation Schedule or Planning for verification of routine cleaning
– Include areas needing periodic cleaning such as coolers and storage facilities.

7° GMP's for Storing & Transport

Good Storing Practices:

Good Storage Practices are valid not only for manufacturers of food products, but also for food and feed importers, exporters, contractors, distributors, wholesalers and retailers. Even at home, as we propose in our courses!

Guideline requirements specified:

- Precautions must be taken to prevent unauthorised persons from entering storage areas.
- Raw materials and food products must not be stored directly on the floor.
- There should be written programmes for many work routines (sanitation, pest control, measures in case of spillage, and cleaning procedures for the sampling area, handling of returned food items).
- Methodic systems used for storage administration (quarantine and storage of rejected materials) have to be validated.
- Maintain the traditional term *"First in – First out"* (FIFO) with the more specific *"first expired/first out"* (FEFO) principle in case you produce relative long shelve life food (canned, bottled, frozen, tetrapack pasteurized/ tyndallized/ uperization drinks or packaged food).
- In addition to the usual GMP requirements on the monitoring of storage conditions (temperature recording with calibrated equipment), temperature mapping should prove the uniformity of the temperature.
- Returned goods may only be returned to saleable stock after the quality has been re-evaluated.

- As for transport, especially the use of ultra frozen ambient in cold chains is discussed and the monitoring of transport conditions with the help of appropriate recording devices is needed like temperature data logger and thermometers attached to storages, tunnels.

All food businesses have to store food, from raw materials or prior to food service; when the food is maintained in the best conditions, at the correct temperature for the appropriate period helps to:

- Prevent food-borne illness

- Preserve the food's taste, appearance and nutritional value

- Provide adequate supplies when they are needed

- Avoid spoilage and wasted food

- Keep within the law and avoid prosecution for selling unfit food.

The correct storage of food can help you and your business to serve safe food, reduce waste and improve product consistency and yields.

Shelving must be stable and durable and maximise all available storage space. Storage containers should protect contents and minimise unnecessary handling. Food storage areas should be easy to clean frequently.

Typical storage areas include:

- <u>Dry Goods Stores</u> – for short and long-term storage of canned and bottled food, cereal, teas, spices, grains

- <u>Refrigerators and cold stores</u> – for storing high risk and perishable foods for short periods

- <u>Chiller cabinets</u> – for food display over short periods

- <u>Freezers</u> – for storage of food for longer periods

- In addition there should be adequate storage made available for cleaning chemicals and equipment housed well away from food storage.

Best Practices in Food Storage

☐ Store products off the floor using dunnage racks and or shelving systems.

☐ Maintain freezers at -18°c or below.

☐ Keep all raw and cooked foods stored separately to prevent cross contamination. Ready to eat foods should be stored above raw foods and both should be air tight covered.

☐ Maintain <u>separate storage sections for fish, meats, dairy produce</u>.

☐ <u>Cover, label and date all food</u> stored in containers.

☐ Use the First in First Out (<u>FIFO) rule</u>.

☐ Use food safe labelling; <u>sticky residue from some materials harbour harmful bacteria and can cross contaminate food</u> contact surfaces of clean containers when nested during storage.

☐ Do not use products that smell bad, or have expired use by date.

☐ Do <u>not overload refrigerators</u> or freezers.

☐ Keep dry store areas well ventilated and food should be cool and dry.

☐ Use food grade food containers when repacking dry goods, air tight lids, labelled and date marked.

☐ Properly store all perishable products within 15 minutes of being unloaded and inspected to ensure safe food temperatures.

☐ Keep all food products covered and stored off the floor on shelves or pallets.

Dry storage of packed cereal grains in India

Good Transportation Practices:

All incoming materials (food and non-food) and finished products are transported, received/shipped, stored and handled under conditions that prevent, eliminate or reduce damage and/or contamination.

Guidelines for carriers

The food company owner or manager (you) should verify that all food carriers are suitable for transporting food. For example:

The temperature during transportation should be controlled to prevent product deterioration.

Adequate cleaning and sanitizing programs should be in place (such as cleaning certificates, wash tickets, Letter of Guarantee and/or record of the previous material transported prior to loading or unloading).

Procedures should be in place to ensure carriers are cleaned adequately and are free from contamination (verified by the operator by conducting visual inspection before loading and upon receipt, by sensory evaluation of ingredients and/or by analysis, as appropriate).

Where the same carriers are used for different food products, cleaning and sanitizing procedures should be in place to prevent cross-contamination of the food (for example, raw versus cooked or ready-to eat food, allergens in products with no allergens).

Where the same carriers are used for food and non-food products, whether in the same shipment or not, procedures should be in place to restrict the transporting of non-food products that can pose a risk to the food products being transported.

For bulk carriers, the operator should have additional cleaning and sanitizing procedures in place. For example:

Bulk tanks should be designed and constructed to permit complete drainage and to prevent contamination. They should be designated to transport a specific commodity and be used for this purpose only.

Cleaning criteria should include the condition of hoses, pumps, inlets, outlets and seals, where applicable.

Where direct contact with food may occur, <u>materials used in carrier construction should be suitable for food contact.</u>

Carriers should be loaded, arranged, and unloaded in a manner that prevents damage and contamination of the food and/or food packaging material.

Sanitizing the transport vehicle for live animals directed to a slaughterhouse in Russia.

Incoming food materials and finished products

Operators should have procedures in place to confirm that incoming food materials meet all purchasing and other documented specifications (organoleptical inspection upon receipt, certificates of analysis, review of labels for allergens, and approved suppliers list).
For <u>imported ingredients</u> and/or products, operators should verify that the suppliers are capable of providing <u>food products that comply with the national legislation of your country</u>.
Incoming materials should always be received and stored in appropriate areas separate from processing areas and finished products (for example, ready-to-eat products).
Products should be stored and handled (<u>proper hands hygiene</u>!) to prevent contamination (for example, microbial growth due to high temperature, rusting of cans) and damage (for example, forklift damage, stacking heights should be controlled).
Written procedures should be available in place to ensure that ingredients stored in opened packages are not contaminated.
Products that are sensitive to environmental conditions (humidity, air, temperature, light) should be stored in appropriate conditions to prevent deterioration: humidity/temperature controlers, cover to prevent disication and/or direct light
A stock rotation procedure should be implemented to minimize deterioration and spoilage (use the "first in, first out": <u>FIFO principle</u>).
<u>Ingredients containing allergens should be clearly identified and stored to prevent cross-contamination with ingredients and products not containing allergens and with other materials and products.</u>

Returned, non-conforming or suspect products should be controlled, clearly identified and segregated in a designated area until they can be disposed of appropriately.

Food packaging materials

The responsible or manager/owner at your food company should inspect food packaging materials before use to prevent using non permitted materials, or damaged, defective or contaminated packaging, which may lead to contamination of the product.

The operator should have the effective written procedures available in place to confirm, always, that contaminated, damaged or defective reusable containers are properly cleaned and sanitized, repaired or replaced, as appropriate.
The operator should have controls in place to prevent contamination of packaging (by <u>pesticides products and non-food chemicals like soaps and sanitizers</u>) and to confirm that packaging material is used for its intended purpose only.

Non-food chemicals

Non-food chemicals should be received and stored in a designated, separated, dry and well-ventilated area. This area should be separate from all food processing, storage, distribution and handling areas (for example, in a separate storage room), so that <u>no possibility exists</u> for cross-contamination of food, food contact surfaces or packaging materials.
<u>Chemicals should be stored and mixed in clean, correctly-labelled containers and dispensed by trained, authorized personnel.</u>

Temperature control

Temperature is controlled appropriately during transportation, handling and storage of food to minimize deterioration of the product.

Ingredients and products should be transported, handled and stored at appropriate temperatures that minimize deterioration, such as microorganism spoilage and rusting of cans.
Ingredients and products requiring <u>refrigeration should be transported</u> and stored at an appropriate temperature (<u>4°C or less but not frozen</u>).
<u>Frozen ingredients and products</u> should be transported and stored at temperatures which do not permit thawing (<u>below 0°C</u>).
Temperatures of food handling areas should be appropriate to the type of products handled and controlled to prevent product deterioration (processing fresh cut vegetables should be done in a refrigerated environment).

Temperatures should be appropriately monitored with proper temperature recording devices. In food transport industry the <u>prevention of food safety problems</u> during transport could be ameillorated if we highlight the risks of physical, chemical, or biological contamination for food during transport:

- Improper refrigeration or temperature control of food products (abuse by excess or lack of cold, frozen or ultrafrozen temperatures).
- Improper management of transportation units (or storage facilities used during transport) to preclude cross-contamination, including improper sanitation, backhauling hazardous materials, not maintaining tanker wash records, improper disposal of wastewater, etc.
- Improper packing of transportation units (or storage facilities used during transport), including incorrect use of packing materials and poor pallet quality;
- Improper loading practices, conditions, or equipment, including improper sanitation of loading equipment, not using dedicated units where appropriate, inappropriate loading patterns, and transporting mixed loads that increase the risk for cross-contamination;
- Improper unloading practices, conditions, or equipment, including improper sanitation of equipment and leaving raw materials on loading docks after hours;
- Poor pest control in transportation units (or storage facilities used during transport);
- Lack of driver/employee training and/or supervisor/manager/owner knowledge of food safety and/or security;
- Poor transportation unit design and construction;
- Inadequate preventive maintenance for transportation units (or storage facilities used during transport), resulting in roof leaks, gaps in doors, and dripping condensation or ice accumulations;
- **Poor employee hygiene**;
- Inadequate policies for the safe and/or secure transport (or storage during transport) of foods, e.g., lack of or improper use of security seals;
- Improper handling and tracking of rejected loads and salvaged, reworked, and returned products or products destined for disposal; and
- Improper holding practices for food products awaiting shipment or inspection, including unattended product, delayed holding of product, shipping of product while in quarantine, and poor rotation and throughput.

To prevent some problems cited above, we recommend:

- Appropriate temperature control during transport;
- Sanitation, including:
  - Monitoring and ensuring the sanitation and condition of transportation vehicles as appropriate;
  - Pest control; and
  - Sanitation associated with loading/unloading procedures;
- Appropriate packaging/packing of food products and transportation units (e.g., good quality pallets, correct use of packing materials);
- Good communications between shipper, transporter and receiver; and
- **Employee awareness and training**.

Bathrooms, Restrooms/Toilets

8° GMP's for Bathroms, Restroms/Toilets

We use to say in every training A6 course on GMP or Food Hygienic Handling that bathrooms must be maintained as clean and sanitized as food plant process facilities and workers hands!

Why? because in this place we look exactly for a better hygiene in our hands, body, skin, face and it is of the most important to keep it properly at any time of day, week or month. If unsafe and not cleanned bathrooms exist, the probabilities of get dirty, contaminate our hands and uniforms are very high, thus elevating the risks of spoil or harm the food we are preparing, manufactoring or storing.

How we can achieve this?

First, the materials used in bathrooms for indoor floor, walls, and ceiling surfaces under conditions of normal use shall be: i) smooth, durable, and easily cleanable for areas, ii) Closely woven and easily cleanable carpet for carpeted areas; and iii) Nonabsorbent for areas subject to moisture such as toilet rooms, and areas subject to flushing or spray cleaning methods.

Cleanability

Floors, Walls and Ceilings.

Except for antislip floor coverings, floors, floor coverings, walls, wall coverings, and ceilings shall be constructed so they are smooth and easily cleanable.

i) Utility service lines and pipes should not be exposed to avoid niches of potential contamination hotspots or sources.

ii) Exposed utility service lines and pipes must be designed to not obstruct cleaning of floors, walls, or ceilings.

iii) Exposed horizontal utility service lines and pipes may not be installed on the floor!

Junctures between Floor and Wall.

Coved, and Enclosed or Sealed.

i) In food companies where the cleaning methods (other than water flushing) are used for cleaning floors, the floor and wall junctures shall be coved and closed to no larger than 1 mm.

ii) The floors in food companies in which water flush cleaning methods are used shall be provided with adequate drains and be graded to drain, with the floor and wall junctures shall be coved and sealed.

Toilets Floor Carpeting Restrictions

A floor covering such as carpeting or similar material may not be installed as a floor covering in: warewashing areas, toilet room areas, handwashing lavatories, toilets, and urinals.

Coverings and Coatings of Wall and Ceiling

i) Wall and ceiling covering materials shall be very well attached so that they are easily cleanable.

ii) Only except in areas used for dry storage, concrete, porous blocks, or bricks used for indoor wall construction shall be finished and sealed to get a smooth, nonabsorbent, easily cleanable surface.

Attachments in Walls and Ceilings

All attachments to toilet's walls and ceilings such as light fixtures, mechanical room ventilation system components, vent covers, towel paper rolls, hand gel soap equipment, wall mounted fans, decorative items, and other attachments shall be of such a desing to ensure an easily cleaning and sanitizing process.

Protective Shielding for Light Bulbs.

Light bulbs must be always shielded, coated, or otherwise shatter-resistant in toilets areas as well as in exposed food areas.

Toilets Insect Control Devices, Design and Installation.

Insect control devices that are used to electrocute or stun flying insects shall be designed to retain the insect within the device.

Toilet Rooms, Enclosed.

Only except where the toilet room is located outside the food company and does not open directly into the food facilities, a toilet room located on the proximities of food handling or cooking facilities must be completely enclosed and provided with a tight-fitting and self-closing door.

Availability of Handwashing Cleansers

Every handwashing sink should be provided with a closely supply of hand cleaning liquid, powder, or bar soap.

Provision of Hand Dryers

Every handwashing sink should be provided with one of these options:

i) Individual, disposable towels

ii) A continuous towel system that supplies the user with a clean towel

iii) A heated-air hand drying device

iv) A <u>hand drying device that employs an air-knife system that delivers high velocity, pressurized air at ambient temperatures</u>

Handwashing Posters.

The poster that notifies food workers how to wash their hands properly should be provided, clearly visible, at every handwashing sinks used by all personnel.

Disposable Towels, Waste Receptacle.

Every handwashing sink that is provided with disposable towels shall be provided with a waste receptacle as specified under.

Minimum Number for Toilets and Urinals

Toilets and urinals shall be provided in a proper proportion to surface and size group of personnel in each area of work in a food company. Ther are many ratios recommended elsewhere, please check it out.

A6 propose 01 toilet each 10 workers/01 urinal each 20 workers.

Not the toilet we want in our food business!

Bathrooms Ventilation

If necessary to keep bathrooms free of excessive heat, steam, condensation, vapors, obnoxious odors, smoke, and fumes, mechanical ventilation of sufficient capacity shall be provided, designed always to extract the air of bathrooms and its potential odors and contaminated spray, dust or gases outside the food company and food handling facilities of course.

Dressing Areas and Lockers

i) Dressing rooms areas must be placed workers change their clothes in the establishment.

ii) Lockers shall be placed in dressing rooms for the orderly storage of workers's clothing and other possessions.

Just before enter to work, please change your clothes by proper food handling uniforms.

- Residues Management

9° GMP's for Residues Management

Again, in all A6 courses given since 1999, we always propose (as for bathrooms!) to get every waste container as clean and sanitized as the food facilities or worker's hands!

Why? well, if a small food company with short worker team number, where one individual should carried out at the same day, be the food handling itself of for example frozen shrimp, and also the vigilance, discharge, cleanning and sanitizing of these very important devices (garbage containers!), the rigourous performance and serious, proffesional attitude to treat, clean and sanitize the waste/garbage containers (and the areas surronding it!) will signify a 99% of reduction of cross contamination risk, if always the mentioned individual wash and sanitize his/her hands properly each time he or she change of role at the small food company.

Even in medium to larger food companies this philosophy should be carried out, preventing to overload these containers, discharging the garbage/waste properly, maintaining its integrity, and performing its cleaning and sanitizing as if food equipment be.

For Waste Disposal the GMP establishes:

-Food companies should operate rigorous systems for waste treatment and disposal in an adequate manner so that they do not constitute a source of contamination in areas where food is exposed.

-Waste material must be properly collected, stored and disposed of in order to inhibit the attraction of rodents and other pests.

-In some processes with food products, more than a half of the raw materials that are used is generated as waste and this will provide two essential conditions for pests to develop: food and water.

-All waste, be liquid or solid, should be stored in closed containers that are not next to open entries into the plant. <u>Waste containers and the areas around them must also be kept clean to eliminate food that can attract pests</u>. These containers and the areas around them should also drain properly to eliminate water that can attract pests.

- Evaluate your system for handling, storing and discarding food waste and other garbage to make sure that it minimizes the potential for contamination and does not attract pests.
- Maintain, clean and sanitize waste receptacles, dumpsters and other elements of your waste system to minimize odor and eliminate breeding places or attractants for pests.

-Monitor your waste disposal, handling and storage system including food waste containers and storage areas, dumpsters and other areas on a routine basis.

How To Monitor

-Monitor your waste control system often to ensure that containers, dumpsters or other waste receptacles have been handled, stored, and emptied in a way to avoid attracting pests. Monitor this at least once a day in your food company, the best possible time might be at the end of the processing day.

-Keep monitoring records; you must keep a record of the results of your observations for your company or official use. Any corrections to problems that could arise should also be written on a monitoring record.

Container with pedal and wheels, excelent choice for your food business!

10º GMP's for Plagues/Pest Management

Pest as rats, mices, birds, insects of varied taxonomic groups, could contaminate food with dangerous microorganisms and/or spoilage microorganisms and thus affect the food safety (and business) and, also could deteriorate the raw food or ready to eat food, the pest itself, if these organisms (macro) develop by feeding on your fruit, cereal or vegetables. Even pest/plagues could contaminate glass containers or cans, externally, and cause illness or death on consumers at the end of food chain, as many examples we use to recall in our courses.

Pests are dangerous, and you must know how to control them properly, rigorously.

Pest Control Program
A complete program that controls pests (Insects, rodents, birds) will fulfill these premises:
1. Prevent entry into facilities.
2. Keep away from them food and water sources.
3. Exterminate as necessary

The vast majority of food companies around the world relys on a licensed/insured PCO (Pest Control Operator) that is recommended always, cause the manage so powerful chemicals that is the better option to be carried out by "outside" personnel that carried out by your own personnel, no matter if big or small food business you are.

The importance of pest control is to prevent pests from contaminating food. Pests can contaminate food through their droppings, hairs, feathers, saliva, faeces, which may carry disease, and also through their urine and through chewing and gnawing on food products or packaging.

To control pests in the plant you must: i) Recognize and report the signs of pest infestation; ii) Remove any conditions that are favourable for pest access and harbourage; iii) Without effective ways to control pests, final products can be affected

Pests need food, water and harbourage to survive. For food manufacturing plants, control of harbourage is the main way of controlling pests. Harbourage may be given through food waste buildup that would protects insects and its larvae, old equipment and material for sealing that is piled unadequetaly around the plant that would harbours pests, and long grass around the plant that hides rodents or vast numbers of trees harboring birds and its nests.

Your food plant will need a written pest control program that will ensure all insects, birds and rodents do not have any opportunity to contaminate, deteriorate or cornsume your food products and ensure that the methods used to control these pests do not cross-contaminate (powerful chemicals) the food products either.

It is mandatory that only qualified individuals are permitted to apply pesticides; these individuales would belong to a third party company specialized in this crucial matter, only that.

Pest Control Device Maps

One very important thing is to develop maps of the whole food comapany architectural plan in 2D or 3D, identifying where pests traps are and where they are caught; this can be used to better understand the problem and prevent further infestations.

In addition to showing where the traps are, the traps should be numbered so that the pest control person can indicate where the pest was found by the number and type of trap. Some examples of pest control devices include:

- Exterior bait stations
- Interior mechanical traps
- Light traps: insects are attracted to lights and then get stuck on a glue board inside the trap
- Pheromone traps: insects are attracted to the pheromone scent and get stuck on a glue board
- In A6 we have propose some clients to control birds (producing a lot of feathers, sloppy feeding debris and semiliquid faecal depositions, all day= risks!) by the use of trained birds (falconry/hawking) to chase pest birds (pigeons, pelicans, seagulls, crows, etc.) that could develop in great numbers if the food plant does not manage properly its garbages, wastes, the cleaning and maintenance of their facilities (both external and internal). This also must be delegated on a third party company specialized in this services with trained birds. Seek around your country.

Permitted Pesticides

Very, very important. Does not matter if your personnel does not "touch" the chemicals involved in killing rodents, insectes or the glues to trap them.

Only approved pesticides can be used in a food manufacturing plant. "Approved" means that the pesticides are approved under your country food laws and Codex Alimentarius list or, for example in this link to Environmental Protection Agency (US):

http://www2.epa.gov/pesticide-registration/registration-information-type-pesticide

Permitted herbicides are very common in our daiy vegetable food items, try to change to organic management each year more please.

Long forbiden DDT insecticide in a majority of countries, but still in use in some others.

Application devices for permitted insecticide products in food business.

68

Rat trap.

A food company should never attempt to perform its own pest control; it is always better and safer, to rely on a third party company with validated experience in the field.

An own (your food comapny) pest control program should be monitoring pests every hour/day/month to ensure that they are kept safely and effectively under control.

There will be an especialized document on this matter, a Standard Operating Procedure (SOP) that should be written down before opening any food business, to define, all the procedures and all responsibilities at your food comapny pest control program, which include planning (you and/or your team) and implementation (third party specialized company!), as well as monitoring (you and/or your team) the efficiency of the program by carrying out regular reviews of collected pest data.

This pest control program must be developed in all areas of your food company: from manufacturing facilities, to storage (raw food, ready to eat food, canned food, packaged food, dry food, etc.) and distribution areas. Also in the storage of empty containers (glass, cans, boxes, etc.), sealing materials, additives, it must be controlled as if food itself is involved, do not forget ever that these will be in contact soon with your food product.

This Standard Operating Procedure (SOP) can be applied to all food business (industry, restaurant, an organic farm, etc.).

11° GMP's for Chemical Products Control

A "chemical" in a food company is any dangerous not edible synthetic substance used inside any food company, like pesticides, greases, lubricants, soaps, detergents, fuel, paints, sanitizers, cooking gases, freezing and cooling gases, even if these not represent a mortal threat to human health.

Because, anyone of our readers could say "but who is going to poison with soap?", well, as we insist in our courses, when you manage a food business you want total quality, not only preserve client's life (the most important of course) but to offer the best product and gain fidelity in consumers around your food business: think there is a drop of liquid soap reach your best chicken soup at your first year open restaurant, what could happens? i) the client does not realize of that taste, but you know a little bit of soap is in his/her soup! (ethical battle?!), ii) your client conplaint loudly that there is a bubble of soap and his/her soup taste like soap! (your sells could star to go down...).

So, if you cary out a food business, the best of the world, you must keep your mind 100% focused in total quality, the perfection of your water glasses surfaces, as clean as diamonds, with no "hair" in your served rices, but talking about "chemicals" do not let them enter in contact with food, or food packaging material or food processing areas, never.

You must develop a **Chemical Control Program**

Cleaners, sanitizers, lubricants, pesticides, and all other non-food chemicals, must be:

-Properly labeled, in all containers.
-Properly stored- away from food or food packaging material or food processing areas.
-Properly used as indicated in the directions given by manufacturers of these chemicals.
-Put away food and packaging or cover before cleaning or using chemicals.

The consumers trust that the food industries linked typical chemical products used are safe, innocuous, effective and of best quality.

Food business must achieve those expectations by applying the GMP consistent manufacturing processes under adequate, rigorous, serious supervision with effective quality control.

It is recommended the use of food grade lubricants to be used in food and beverage manufacturing plants from the time that raw materials arrive until after final packaging to improve food safety.

A majority of food and beverage manufacturers in the World are still using non-food grade oils and greases when making food or beverage products.

After the U.S. Department of Agriculture (USDA) and the U.S. Food and Drug Administration (FDA) assessed and organized the usage of food-grade lubricants. The primary current food-grade lubricant classifications are:

- **H1:** Lubricants used in food processing environments where there is the possibility of incidental food contact.
- **H2:** Non-food-grade lubricants used on equipment and machine parts in locations where there is no possibility of contact.
- **H3:** Food-grade lubricants, typically edible oils, used to prevent rust on hooks, trolleys and similar equipment.
- **21.CFR 178.3570:** Outlines allowed ingredients for the manufacture of H1 lubricants
- **21.CFR 178.3620:** White mineral oil as a component of non-food articles intended for use in contact with food
- **21.CFR 172.878:** USP mineral oil for direct contact with food
- **21.CFR 172.882:** Synthetic iso-paraffinic hydrocarbons
- **21.CFR 182:** Substances generally recognized as safe.

About the formulation requirements for food grade lubricants and greases, the minimum standard states that these products must not contain added heavy metals, and shall not contain ingredients classified as carcinogens, mutagens or teratogens. Also it is recommended that this useful and important chemicals in food industries machines be neutral in taste and odor, and resistent to temporal, chemical, biological, thermal or mechanical stresses without rapid degradation or changing its neutral state.

In the case of soaps & detergents + sanitizers, these must meet the cleaning and disinfection objecives that they are designed for, but at the same time to not be an excessive potent substance to harm human health in case of accidental cross contamination. Spo written procedures shall be implemented for approval, storage and use of approved cleaning and sanitizing chemicals. Chemical products like these shall not be used unless approved for the specified applications by a technically qualified person and from a reliable manufacturer.

The soaps & detergents & sanitizers you might use could be same you might use in at home. Try to use eco friendly soaps & detergents (which are less bubble maker, a common feature associated with "more efective soaps" that will reduce your pollution rates and if you communicate this to your clients, you will get the Eco label from society at least in this subject of avoiding excess loads of phosphorus and/or nitrogen inputs into aquatic systems).

For sanitizers, keep all this products in your food company, in order of using it alternatively and avoid adaptation of microbes to only one of these products.

i) Hypochlorites

The effectiveness, low cost and ease of manufacturing make hypochlorites the most widely used sanitizers. Sodium hypochlorite is the most common compound and is an ideal sanitizer, as it is a strong oxidizer.

Sodium hypochlorite can be available easily in the Internet.

Hypochlorites cause broad microbial mortality. These compounds inhibit cellular enzymes and destroy cell membranes and DNA moleculesThe spores, however, are more resistant to hypochlorites, as the spore coat is not susceptible to oxidation except at high concentrations coupled with long contact times at elevated temperatures.

While hypochlorites are very reactive, their useful properties are negatively impacted by factors such as suspended solids, high temperatures, light, water impurities and improper pH levels. In routine use, surfaces must be as free as possible of organic materials, and the pH must be maintained between 5 to 7 to ensure that the greatest amount of hypochlorous acid is available. As with any sanitizer, measurements must be taken periodically to make certain that the freely available chlorine is at the desired level. For no-rinse applications, the maximum allowable concentration (by food related laws) of available chlorine is 200 ppm.

Other disadvantages of hypochlorites are corrosiveness to metals, health concerns related to skin irritation and mucous membrane damage and environmental contamination.

ii) Chlorine Dioxide

This inorganic compound is a broad sanitizer effective against bacteria, fungi and viruses. Chlorine dioxide is an oxidizer that reacts with the proteins and fatty acids within the cell membrane, resulting in loss of permeability control and disruption of protein synthesis.

Remember that chlorine dioxide is an explosive gas, but it is relatively safe in solution. It is produced on-site as it can't be compressed or stored commercially in gaseous form.

Compared with hypochlorites, chlorine dioxide requires much lower concentrations to achieve microbial mortality. For example, a 5-ppm solution is effective as a sanitizer on food contact surfaces with a contact time of at least 1 minute. Further, disinfection can be achieved with 100 ppm using a contact time of 10 minutes.

Chlorine dioxide functions well over a pH range of about 6 to 10, thus allowing increased mortality of some microbes at higher values. Another advantage is that chlorine dioxide does not form chlorinated organic compounds, making it more environmentally friendly.

iii) Iodophors

These compounds are less active than hypochlorites but are effective sanitizers. Iodophors attach to the sulfurs of proteins such as cysteine, causing inactivation and cell wall damage. Carriers with iodophor solutions allow a sustained-release effect, resulting in continuous microbial mortality.

Iodophors fare better in situations in which the pH is slightly acidic (just below 7), as less active forms exist above neutral pH. The common concentration for sanitization is 25 ppm for 1 minute. Unfortunately, iodine compounds easily stain many surfaces, particularly plastics. On the positive side, they are common sanitizers used on glass surfaces, such as in the beer and wine bottling industries. The USEPA has assessed iodophors as having no significant effect on the environment.

iv) Peroxyacetic Acid (PAA)

<u>PAA is an effective sanitizer that is active against many microorganisms and their spores</u>. Mortality is produced by the disruption of chemical bonds within the cell membrane. PAA-based sanitizers are frequently paired with stabilized hydrogen peroxide. These sanitizers function well under cold conditions (~ 4 °C), thus producing acceptable microbial mortality on equipment normally held below ambient temperature. PAA is also effective in removing biofilms and is more active than hypochlorites.

PAA solutions can be attenuated by the organic load and will begin to lose activity as the pH approaches neutral. These solutions are applied at concentrations ranging from about 100 ppm to 200 ppm for peroxyacetic acid and 80 ppm to 600 ppm for hydrogen peroxide.

PAA-based sanitizers are environmentally friendly as the compounds therein break down into acetic acid, oxygen and water. These sanitizers are also less corrosive to equipment than hypochlorites. As with any highly active oxidizer, concentrated PAA can present a safety hazard.

v) Quaternary Ammonium Compounds (QACs or Quats)

Quaternary ammonium compounds are fairly complex chemicals in which nitrogen is bound to four organic groups. In general, QACs are effective against a wide range of microbes cell membranes, although the spore phase is unaffected. At lower concentrations, Gram-positive bacteria are more sensitive to QACs than Gram-negative bacteria.

QACs may be applied at concentrations varying from about 100 ppm to 400 ppm. As sanitizers, QACs are commonly applied at 200 ppm to food contact surfaces, and the solution is allowed to dry. Once dry, a residue of the QAC compounds remains and provides germicidal activity until degradation occurs.

QACs are usually odorless, nonstaining, noncorrosive and relatively nontoxic to users. They function well over a broad temperature range and a wide pH range, although activity is greater at warmer temperatures and in alkaline situations.

vi) "Artisanal choices"

If you have a small food business, you might apply, on your stainless steel tables, tools, service glasses, dishes, etc., some of these smart and cheap substances: citric acid (oneself pasteurized and concentrated lemon juice), acetic acid (oneself pasteurized vinegar), brine solution (oneself pasteurized table salt solution in clean filtered water) or isopropilic alcohol (as used in microbiology labs, controling a flame on clean and dry surfaces after application of earlier "artisanal choices" during sanitizing). All these products your might produce it yourself, and must be applied with new or very clean + sanitized cloth, towel or rag and rinsed with clear, odorless, clean tap water.

Controlled alcohol flame on food handling surface tables after the cleaning+sanitizing process. A possible 100% sterile environment to work then. It is a quick way to sterilize glass or stainless steel utensils too. This method needs a extreme caution, cause a burning drop of alcohol can easily start a fire if dropped on something flammable (paper towels, clothes).

About the pesticides, the USEPA sets thees categories for pesticides:

i) Conventional pesticides, includes all ingredients to control rodents and insects, other than biological pesticides and antimicrobial pesticides.

ii) Antimicrobial pesticides, substances or mixtures of substances used to destroy or supress the growth of harmful microorganisms whether bacteria, viruses, or fungi on inanimate objects and surfaces.

iii) Biopesticides, types of pesticides derived from certain natural materials.

iv) Inert ingredients, substances contained in pesticides in addition to the active ingredient(s)

All theses "chemicals", the cleaning + sanitizing ones, the lubricants and greases, and the pesticides (insecticides and rodenticides) and other non-food-ingredient chemicals must be stored in a designated area segregated from the areas used for storage of raw materials and products and away from process facilities.

12º GMP's for Glass and Foreign Material Control

Physical dangers are inhereted to any food business activity, and represent hazards to human health by possible injuries in mouth, lips, teeth, tongue, throat or suffocation caused by solid materials that could reach the food we are producing, selling or transporting.

I have founded several times in slowfood and fastfood restaurants: wedding rings, toohpicks, hairs, paper, etc.

Even if I or my family got no injure at all, those details tells us something about these food business: they (cooker, manager, owner) simply do not care in Total Quality, meaning that someone in the kitchen should verify and control all aspects involved in a good service, safe, inocuous, trustable food products you offer, not only serving or cooking temperatures, but visually check the perfect state of the food product, by your own eyes, or in a big industry with optical hardware devices (laser, infrared) like those in brewery factories for bottled beers or wine to avoid foreign materials be inside the bottles once is close the cap or cork. Because this is a recurrent risk in food business we have to be very focused in preventing such situations with our products.

Optical devices to prevent foreign material on bottled drinks. Eyes supervision could works too in smaller food businesses.

Develop a written Program to Control Foreign Material debris

Program to control glass and glass-like (brittle plastic) foreign material.

Please count and place in a map of your food business:

i) Cover glass light bulbs
ii) Monitor glass packaging and other glass in facility
iii) Try to prevent or at least detect breakage

Roof lights covering, prevents possible breakage and cancallation of a whole food batch or lot.

Program to control Foreign Material Sources

i) Metal
ii) Nuts, bolts, mixers, knives, thermometers, grinders
iii) Wood
iv) Crates, pallets
v) Plastic
vi) Box bands, packaging, broom handles
• Hair, gum, jewellery, pens, cigarette butts, etc.

Other types of foreing material that could reach our food products. Also, these can cause malfunction of food machinery if accidentally reach the food automated processes.

Risks

i) Poor <u>personal commitment on safety behavior</u> (lack or little efficacy of training courses).

ii) Little or no investments on automated devices to detect foreign material in food products.

iii) Little or no efforts to control visually by personnel or managers/owners.

What to do in foreign material control

i) Visual inspection
– Raw materials during receiving
– In process- carefully add ingredients to products
– Final product/ outgoing food
– watch for potential contaminants

ii) Good personal hygiene and training

iii) Good housekeeping/cleaning practices

X-ray inspection will be routine in next year for big food industries

13° GMP's for Food Defence/Prevention of Intentional Food Contamination

Food defence/ Security- Prevention of intentional contamination of food.

The rule on food defense is, to require domestic and foreign facilities to address vulnerable processes in their operations to prevent acts on the food supply intended to cause large-scale public harm. This is a new modern requirement from the FDA Food Safety Modernization Act, where it is demanded to biggest food businesses to develop a written food defense plan that copes with possible and significant vulnerabilities in any food product and the operations needed to manufacture it.

The proposed rule is: to protect food products, ingredients, additives, from human intentional adulteration whit the intention of causing large-scale public harm or fatalities. There are 4 main procedures in the food systems that are very vulnerable to adulteration.

- i) bulk liquid receiving and loading;
- ii) liquid storage and handling;
- iii) secondary ingredient handling (the step where ingredients other than the primary ingredient of the food are handled before being combined with the primary ingredient); and
- iv) mixing and similar activities.

Raw milk reception

Some facilities are required to check the production system to determine if it has any of these former procedures types and complete our own vulnerability assessment.

WHO issued its "Terrorist Threats to Food—Guidelines for Establishing and Strengthening Prevention and Response Systems" to provide a policy guidance to its Member States for integrating consideration of deliberate acts of sabotage of food into existing prevention and response programs.

WHO uses the term "food terrorism" and defines it as "an act or threat of deliberate contamination of food for human consumption with biological, chemical and physical agents or radionuclear materials for the purpose of causing injury or death to civilian populations and/or disrupting social, economic or political stability."

Focusing on the two key strategies of prevention and response, WHO recommends that all segments of the food industry consider the development of security and response plans for their establishments, proportional to the threat and their resources. The guidelines state that the key to preventing food terrorism is enhancing existing food safety programs and implementing reasonable security measures on the basis of vulnerability assessments. The guidelines further state that the most vulnerable foods, food ingredients, and food processes should be identified, including: the most readily accessible food processes; foods that are most vulnerable to undetected tampering; foods that are the most widely disseminated or spread; and the least supervised food production areas and processes. These are guidelines to assist the food industries in protecting food against intentional adulteration

To begin, get the following documents:

- A labeled map of the facility
- All written operational procedures, such as HACCP, GAP, BMP, GMP and SOPs
- Procedures related to your workforce, such as preemployment screening, psicological tests, food hygiene trainning and food security training.

14° GMP's for Allergens Ingredients Control

Food allergies have become a great concern for any food manufacturer, importer, and distributor. <u>Increased consumer awareness,</u> together with the recognition of the serious consequences of undeclared allergens, and, of course, the economic impact of food recalls has all provoked a public debate today. Food business must take steps to prevent the presence of undeclared ingredients in their products that may cause allergic reactions in sensitive individuals.

The Food Quality Plans like Hazard Analysis Critical Control Point (HACCP), <u>Good Manufacturing Practices (GMPs)</u>, and Good Importing Practices (GIPs) are food safety controls with respect to preventing allergen contamination. Some food businesses have created successful allergen prevention plans which <u>include allergy awareness of all staff and company commitment.</u>

Every food business producing, transforming, transporting, selling or preparing food for human comsuption must develop some minimal measures to avoid allergens. The major allergens include: <u>peanuts, tree nuts, milk, wheat, soy, fish, crustacean shellfish, and eggs.</u>

If your food plant or restaurant, to deal with these products you must carry out an Allergen Control Program, which consist in assure that:

• <u>Raw materials containing allergens are properly labelled and stored.</u>
• Movement of allergen- containing ingredients within facility is controlled to not cause cross contamination.
• Sanitation procedures to ensure <u>complete removal of allergens</u> from equipment.
• Procedures to ensure the <u>labelling of finished product</u> and rework containing allergens.

Food business owner, manager or the responsible person must be aware that some actions taken by individuals, be personnel, operators or staff have to be trained to avoid, detect and inform others in these subjects:

i) avoid cross contact with foods containing allergens

ii) prevent cross contact with unlabeled allergenic material

iii) adopt allergen control practices within the facilities

iv) abbility to achieve the identification of ingredients containing food allergen(s); the of these ingredients (*e.g.*, physical segregation); process controls; verified cleaning processes; label controls and label review; and **employee training**."

v) commitment in the continuous revision of current GMPs to include guidelines regarding rework and shared equipment; guidance on the need for employee training regarding food allergies; and guidance on the use of precautionary ('may contain') statements.

Remember that undeclared allergens and gluten sources could represent a serious or life threatening health risk for allergic or sensitive individuals. Also, undeclared gluten may contribute to chronic health issues for those individuals with Celiac disease or gluten sensitivity.

Again, remember that the main causes of undeclared allergens in foods.

- Cross-Contamination or Carry Over
- Inapropriate Use of Rework
- Ingredient Changes, Substitutions or Additions
- Incorrect Labels
- Incorrect or Incomplete List of Ingredients
- Unknown Ingredients
- Misrepresentation of Common Names and Hidden Allergens
- Labelling Exemptions

Controlling food allergens in the food business is part of GMPs

An Allergen Prevention Plan (APP)

Even if systems such as GMP's and HACCP are employed, an allergen prevention plan (APP) must be implemented for any operation in which allergen sources exist in order to effectively manage food allergy risks. To develop an effective APP, led by an allergy-prevention team, we have to make a risk evaluation, an allergen mapping, an ingredient control, engineering and system design, traffic patterns, work-in-progress, rework, maintenance, packaging and labelling and training*.

All these steps are needed to be undertaken rigourously to ensure that even the small amounts of common allergens are addressed. *The most important is to educate, continuously, our food plant staff/workers.

Developing Your Allergen Prevention Plan

Allergens are a major issue for many food manufacturers, packers, importers, distributors and officials. The ameillorated consumer awareness, the improvements in allergen detection methods, recognition of the serious consequences of undeclared allergens, and the financial impact of food recalls have all served to raise the visibility of this crucial matter. It is important that any size food bussiness take steps to prevent the presence of undeclared ingredients in their products that may cause allergic reactions in sensitive individuals.

Hazard Analysis Critical Control Point (HACCP) systems, Good Manufacturing Practices (GMP's) and Good Importing Practices (GIP's) are food safety control plans with respect to preventing allergen contamination.

Effective steps can include the following:

1. Establish an allergy control team: In small food plants it is recommended to have at least one person who has the responsibility for assessing and maintaining an allergy prevention plan.

2. Identify the key allergens causing sensitivities and intolerances in your food products:

The most dangerous food reactions are caused by some natural organic molecules like protein, modified protein, lipo-protein fractions and fatty acids, but there are dozens or molecules affecting our inmune system, antibodies, even inorganic molecules as sulphites or chlorine.

- Yeast (*Saccharomyces cerevisae*) and yeast related food (bread, alcohol)
- Peanuts (Ara peptides: short proteins; Arachidonic acid: a lipid with inflammatory lipidic mediators involved in the pathogenesis of allergic diseases
- Alcohol molecule itself.
- Tree nuts (proteins founded in almonds, Brazil nuts, cashews, hazelnuts, macadamia nuts, pecans, pine nuts, pistachios, walnuts)
- Sesame seeds (an albumin, not glycosylated)
- Milk (lactose - a natural sugar, disaccharide- intolerance; caseine and almost a dozen of other proteins "floating" in milk)
- Mustard seed (proteins related to tree nuts ones, and pollen raletad molecules too)
- Eggs (albumins and > 40 different proteins from the "white"; vitellenins (2) and phosvitin from the yolk.
- Seafood proteins (fish: parvalbumins, crustaceans and shellfish: tropomyosin)
- Soy (P34, a thiol protease of the papain family)
- Wheat (glutein protein = gluten)
- Gluten sources (barley)
- Biotoxins (Ciguatoxins from shellfish in contact with specific microalgae: dinoflagellates of the genus *Gambierdiscus;* Scombrotoxin from fish of the *Scombridae* and *Scomberesocidae* families, such as mackerel and tuna).
- Anisakis (allergies to a foodborne parasite, *Anisakis simplex*, have been linked to the ingestion of this nematode, which causes infections in humans called Anisakiasis ('herring or cod worm' disease) and induces immune reactions.
- Sulphites (occur naturally in all wines; used as food preservative additive in fruits, shrimps)

- Chlorine (usual sanitizer in process water and in work surfaces and equipments; its abuse can lead to contamination of processed food)

Allergic reactions are severe adverse reactions that occur when the body's immune system overreacts to a particular protein or other molecules like sulphites that may cause severe adverse reactions (Sulphite sensitivity reactions) similar to those with food allergies.

All companies may wish to identify other allergens of specific concern to its products!

Conduct a hazard analysis on the incoming materials and processing steps to determine areas of greatest concern. Each product, process and possible allergen should be assessed. This <u>can be done by preparing a master list of all ingredients, raw materials and products</u>, identifying those that contain allergens, and indicating any special handling and storage instructions for each of them.

Prerequisite Program, GMP's:

- Storage and Handling: Store and handle allergenic foods and ingredients in a manner that will avoid cross-contamination with other food products. This may require dedicated storage areas for the various allergen-containing ingredients handled in your food business. When this latter is not possible, please store allergenic foods and ingredients in separate heights, as being stored below non-allergenic foods and ingredients, (in bottom shelf/rack, to prevent allergens falling on other foods and ingredients). Ensure that allergenic foods and ingredients are clearly identified by signs or colour codes.
- Equipment and System Design: Determine which products and production lines present the highest risk. Consider dedicated lines for certain allergen-containing foods. Minimize the amount of equipment that comes into contact with allergenic ingredients by isolating allergen addition points and adding allergenic ingredients near the end of the process. Reduce cross-over of lines and design equipment to allow for complete cleaning and inspection.
- Cleaning and Maintenance of Equipment: Ensure that your sanitation program addresses the need to remove allergenic foods. Disassemble equipment and manually clean where necessary. With equipment that is very difficult to clean, you may consider dedicating it to specific allergen-containing foods.
- <u>Employee Training</u>: Incorporate allergy awareness and controls into the training for all employees, <u>using a variety of approaches and educational materials</u> to <u>reinforce awareness and commitment.</u>

Review process controls

- Incoming Ingredients: <u>Obtain accurate ingredient information from all suppliers to identify all allergens,</u> including allergens present as components (ingredients of ingredients). Obtain ingredient specification sheets, letter of guarantees, certificate of analysis or a complete list of ingredients.
- Product Formulation: Provide current written formula to production employees. <u>Establish process controls to ensure that multi-component products are produced in accordance with the formula,</u> to ensure a consistent product and prevent the addition of allergen-containing ingredients not listed on the label.
- Allergen Mapping of the Plant: Using flow diagrams of the processes for multiple production lines, <u>identify equipment that is used for both allergenic and non-allergenic products</u>. The identified equipment may have greater potential for allergen cross-contamination.
- Production Scheduling: <u>If dedicated lines are not available, products containing allergens should be scheduled at the end of production runs</u> so equipment can be thoroughly cleaned, before the next run of products. The same precautions should be in place to avoid cross-contamination between products containing different allergens.
- Labelling and Packaging: Ensure that the labelling and packaging is accurate and matches the right product. Ensure that ingredient substitutions are reflected on the label. Dispose of outdated labels.
- Communicate with courses and meetings, the <u>food allergy and sensitivity awareness regularly to employees and suppliers, in continued formation</u>.
- Regularly assess the allergy prevention plan: This is to ensure continual effectiveness. Reassess after any changes are made to products or processes, e.g., ingredient suppliers, formulations, equipment, process flows, production schedules, etc.

15° GMP's for Clients Complaints

TRANSPARENT RESOLUTION PROCESS

Report a Complaint

Begings a constructive feedback with your clients will help your food business improve the quality of products and services it provide. Anyone may submit a complaint.

Food business can not accept anonymous complaints. Be sure that an appropiate method to create complaints on line, by telephone, by postal mail, it is available and with a system that fulfills the minimal information (address, phone number, national identity number or passport, work address) on groups or individuals making the complaint. This information will not be shared with any third parties without the permit of the person or group who provided the information.

How is fulfilled a complaint

Any complaint on your food product does not necesarily need to follow a specific format. But at least must provide the following details:

- Name, address and other contact information such as phone and fax numbers, cell phone and email address.
- Representants of a complainant must provide contact information for itself and the group/person being represented.
- A description of the nature of the complaint
- Background information on the complaint, including the names of your Investment Advisor or branch manager, a chronology of events, and the steps you may already have taken in an attempt to resolve the issue or raise your concerns.

Informing when received

GMP programs will undertakes a response to acknowledge receipt of the valid complaint in up to five business days.

Reviewing the complaint

The review and assessment stage may be completed quickly, or may require further in-depth assessment. Your food company will review the complaint and the circumstances surrounding it. If more information is required, the food company will contact the client at the contact information provided. For last, once it is concluded the review, the food company will notify the client in writing, explaining how its complaint will be resolved.

As stayed by WHO: "All complaints and other information concerning potentially defective products must be carefully reviewed according to written procedures and corrective action should be taken".

Procedures

Designated responsible person or team must: to handle complaint, to decide on measures to be taken, to access to records.

SOP written procedure: describes actions to be taken, among whom includes the need to consider if a recall procedes by possible product defect, contamination or post contamination.

During investigation: Complaint team must invite the Quality Control team of your food company to get involved, in stablishing what was wrong or may have been the cause. This will be executed in a fully recorded investigation that reflects every details of the case. The team involved must check other batches of produuct with suspected or reprocessed products,

Inform authorities: in case of serious quality problems like defects of manufacture, product deterioration, adulteration or counterfeiting.

Inform competent authorities in case of important quality failure like faulty manufacture, product deterioration or adulteration, counterfeiting.

Faulty manufacture is critical defects like incorrect labelling of products or microbiological post contamination of a sterile food product.

Labelling & Traceability/ Documenting/ Archives

16° GMP's for Proper Labelling & Traceability/Documenting/Archives

Labeling – System to ensure proper labeling.

Manufacturers must incorporate in their quality assurance program several elements that relate to the labeling controls in order to meet the Good Manufacturing Practice (GMP) requirements of any Quality System regulation. The quality assurance program must be adequate to ensure that labeling meets the GMP device master record requirements with respect to legibility, adhesion, etc., and ensure that labeling operations are controlled so that correct labeling is always issued and used.

Labeling includes equipment labels, control labels, package labels, directions for use, maintenance manuals, etc. The displays on CRT's and other electronic message panels are considered labeling if instructions, prompts, cautions, and parameter identification information are given.

A good labelling practice must achieve the traceability and recall points of your Food Company, and have to help you or your team in knowing where the ingredients come from, which ingredients are in every batch of product, where is going the batches.

To trace and recall well you will need these elements: an accurately documented receiving and tracking of ingredients through facility (handout)
• Lot coding system
– May be one day of production, one batch, etc.
• Accurately documented distribution
• Plan in place to conduct the recall process,
verified with mock recall.
• Crisis response plan in place.

Documenting and Archives

If it's not written down, then never happened. This logical rule in any good manufacturing practice (GMP) regulation specifies that the food manufacturer must achieve a proper documentation and recordings.

Documentation helps to build up a detailed picture of what a manufacturing function has done in the past and what it is doing now and, thus, it provides a basis for planning what it is going to do in the future. Regulatory inspectors, during their inspections of manufacturing sites, often spend much time examining a company's documents and records. Effective documentation enhances the visibility of the quality assurance system. In light of above facts, we have made an attempt to harmonize different GMP requirements and prepare comprehensive GMP requirements related to 'documentation and records,' followed by a meticulous review of the most influential and frequently referred regulations.

Documentation is the key to GMP compliance and ensures traceability of all development, manufacturing, and testing activities. Documentation provides the route for auditors to assess the overall quality of operations within a company and the final product.

Tips:

- Good documentation constitutes an essential part of the quality assurance system. Clearly written procedures prevent errors resulting from spoken communication, and clear documentation permits tracing of activities performed.
- Documents must be designed, prepared, reviewed, and distributed with care.
- Documents must be approved, signed, and dated by the appropriate competent and authorized persons.
- Documents must have unambiguous contents. The title, nature, and purpose should be clearly stated. They must be laid out in an orderly fashion and be easy to check. Reproduced documents must be clear and legible.
- Documents must be regularly reviewed and kept up-to-date. When a document has been revised, systems must be operated to prevent inadvertent use of superseded documents (e.g., only current documentation should be available for use).
- Documents must not be handwritten; however, where documents require the entry of data, these entries may be made in clear legible handwriting using a suitable indelible medium (i.e., not a pencil). Sufficient space must be provided for such entries.
- Any correction made to a document or record must be signed or initialed and dated; the correction must permit the reading of the original information. Where appropriate, the reason for the correction must be recorded.
- Record must be kept at the time each action is taken and in such a way that all activities concerning the conduct of preclinical studies, clinical trials, and the manufacture and control of products are traceable.
- Storage of critical records must at secure place, with access limited to authorized persons. The storage location must ensure adequate protection from loss, destruction, or falsification, and from damage due to fire, water, etc.
- Records which are critical to regulatory compliance or to support essential business activities must be duplicated on paper, microfilm, or electronically, and stored in a separate, secure location in a separate building from the originals.
- Date may be recorded by electromagnetic or photographic means, but detailed procedures relating to whatever system is adopted must be available. Accuracy of the record should be checked as per the defined procedure. If documentation is handled by electronic data processing methods, only authorized persons should be able to enter or modify data in the computer, access must be restricted by passwords or other means, and entry of critical data must be independently checked.
- It is particularly important that during the period of retention, the data can be rendered legible within an appropriate period of time.
- If data is modified, it must be traceable.

There are various types of procedures that a GMP facility can follow. Given below is a list of the most common types of documents, along with a brief description of each.

1. *Quality manual*: A global company document that describes, in paragraph form, the regulations and/or parts of the regulations that the company is required to follow.
2. *Policies*: Documents that describe in general terms, and not with step-by-step instructions, how specific GMP aspects (such as security, documentation, health, and responsibilities) will be implemented.
3. *Standard operating procedures (SOPs)*: Step-by-step instructions for performing operational tasks or activities.
4. *Batch records*: These documents are typically used and completed by the manufacturing department. Batch records provide step-by-step instructions for production-related tasks and activities, besides including areas on the batch record itself for documenting such tasks.
5. *Test methods*: These documents are typically used and completed by the quality control (QC) department. Test methods provide step-by-step instructions for testing supplies, materials, products, and other production-related tasks and activities, e.g., environmental monitoring of the GMP facility.

    Test methods typically contain forms that have to be filled in at the end of the procedure; this is for documenting the testing and the results of the testing.

6. *Specifications*: Documents that list the requirements that a supply, material, or product must meet before being released for use or sale. The QC department will compare their test results to specifications to determine if they pass the test.
7. *Logbooks*: Bound collection of forms used to document activities. Typically, logbooks are used for documenting the operation, maintenance, and calibration of a piece of equipment. Logbooks are also used to record critical activities, e.g., monitoring of clean rooms, solution preparation, recording of deviation, change controls and its corrective action assignment.

Glosary in documentation

Approved by: signature of a qualified individual (supervisor or designee) indicating that the information documented is complete, accurate, and acceptable.

Backdating: is the practice of going back to a previously completed task that has not been properly initialed and dated and placing the date that the task was completed on the date line, as thorough filling in the date had been done in a timely fashion. This practice is not allowed in any cGMP document.

Batch Production Record: Collection of records associated with the manufacture of a specific lot of product. Record containing all quality-relevant planned and actual data on the production of a batch.

Bundesgesundheitsamt (BGA)
German Federal Health Organization. The German Government agency that must approve new pharmaceutical products for sale within Germany, it is the equivalent of the U.S. Food and Drug Administration (FDA).

cGMP
Current Good Manufacturing Practices. The set of current, up-to-date methodologies, practices, and procedures mandated by the Food and Drug Administration (FDA) which are to be followed in the testing and manufacture of pharmaceuticals. The set of rules and regulations promulgated and enforced by the FDA (through Biannual inspections) to ensure the manufacture of safe clinical supplies. The cGMP guidelines are more fine-tuned and up to date (technologically speaking) than the more general GMP. The cGMP guidelines are going through further modification this year 2003.

Comment: Any written additions to a document for informational purposes. All comments must be initialed and dated by the person writing the addition and may require a verification.

Controlled documents: Written approved documents used in association with cGMP-related activities to ensure compliance with U.S. and international regulations, as wall as company standards.

Cross-out: A cross-out indicated a correction has been made. This is accomplished by drawing a single straight ink line through information which has been entered inadvertently or incorrectly. All cross-outs must be initialed and dated.

Data: the values and information generated by processing, calculating or transcribing from the raw data. This may include computer printouts.

Date: the actual day on which information is entered or printed on a document.

Document: A written or printed form which is used to furnish information or provide instructions.

European Medicines Evaluation Agency (EMEA)

A London-based agency of the European Union (EU) that began operation in 1995. It coordinates drug licensing and safety matters throughout the nations of the EU. Its licensing/approval process is compulsory throughout the EU.

Food and Drug Administration (FDA)

The federal agency charged with approving all pharmaceutical and food ingredient products sold within the United States. FDA homepage

Good Manufacturing Practices (GMP)

The set of general methodologies, practices, and procedures mandated by the Food and Drug Administration (FDA) which is to be followed in the testing and manufacture of pharmaceuticals. The purpose of GMPs is essentially to provide for record keeping and in a wider context to protect the public. GMP guidelines exist instead of specific regulations due to the newness of the technology, and may later be superceded (modified) due to further advances in technology and understanding.

Identifiers: Information that serves to identify or describe something, such as effective dates, lot number, line number, equipment number, manufacturing or task date, product description, container numbers, specification number, run number, Identifiers can usually be retrieved form another source or document.

Initials: consist of the first letter or both the first name and last name(surname). Use of the middle initial is optional but one person should be consistent in how they write out their initials.

Label: on products or solutions should give the contents of the item, when it was created and who is responsible for the product. Will sometimes give batch number and should also give pertinent safety information.

Logbook: contains the records of a performance, a list of how variable change over time especially on pieces of equipment.

NA or N/A : Abbreviation for the phrase "not Applicable." It is used to indicate that the entering of date into a space provide is not appropriate in that particular case.

Overwriting: refers to writing over previously recorded information to make a change. Overwriting is never allowed on any cGMP document.

Process Validation
(for production of a pharmaceutical) Defined by America's Food and Drug Administration (FDA) as "Establishing documented evidence which provides a high degree of assurance that a specific process will consistently produce a (pharmaceutical) product meeting pre-determined specifications and quality characteristics."

Protocol: an original draft or record of a document that plans for a scientific experiment

Performed by: Initials or signature of the person executing an operation or task (usually the "operator" or "analyst").

Quarantine: the default status for raw materials and packaging components upon receipt from he supplier and for drug products upon completion of processing while waiting evaluation against identified release criteria.

Raw Data: the actual information obtain from an observation, test, measurement or activity. This may include computer or instrument printouts.

Recorded by: initials or signature of a person documenting information, results, or readings of an operation (may be the "operator").

Reviewed by: Initials or signature of the person examining a task, document or record in order to confirm its accuracy and completeness, inducing checking calculations.

Signature: consist of a least the initial of the first name and complete last name.

GMP general Glossary

"Adulterant" means any material which is or could be employed for making the food unsafe or sub-standard or misbranded or containing extraneous matter.
"Consumer" means persons and families purchasing and receiving food in order to meet their personal needs.
"Contaminant" means any substance, whether or not added to food, but which is present in such food as a result of the production (including operations carried out in crop husbandry, animal husbandry or veterinary medicine), manufacture, processing, preparation, treatment, packing, packaging, transport or holding of such food or as a result of environmental contamination and does not include insect fragments, rodent hairs and other extraneous matter;
"Food" means any substance, whether processed, partially processed or unprocessed, which is intended for human consumption and includes primary food to the extent defined in clause (zk) genetically modified or engineered food or food containing such ingredients, infant food, packaged drinking water, alcoholic drink, chewing gum, and any substance, including water used into the food during its manufacture, preparation or treatment but does not include any animal feed, live animals unless they are prepared or processed for placing on the market for human consumption, plants, prior to harvesting, drugs and medicinal products, cosmetics, narcotic or psychotropic substances :
"Food Authority" means the Food Safety and Standards Authority of India established under section 4;
"Food business" means any undertaking, whether for profit or not and whether public or private, carrying out any of the activities related to any stage of manufacture, processing, packaging, storage, transportation, distribution of food, import and includes food services, catering services, sale of food or food ingredients;
"Food business operator" in relation to food business means a person by whom the business is carried on or owned and is responsible for ensuring the compliance of this Act, rules and regulations made there under
"Hazard" means a biological, chemical or physical agent in, or condition of, food with the potential to cause an adverse health effect;

"Food safety" means assurance that food is acceptable for human consumption according to its intended use;

"Food safety audit" means a systematic and functionally independent examination of food safety measures adopted by manufacturing units to determine whether such measures and related results meet with objectives of food safety and the claims made in that behalf;

"Food Safety Management System (FSMS)" means the adoption Good Manufacturing Practices, Good Hygienic Practices, Hazard Analysis and Critical Control Point and such other practices as may be specified by regulation, for the food business;

"Food Business Operator" in relation to food business means a person by whom the business is carried on or owned and is responsible for ensuring the compliance of the Act, rules and regulations made thereunder

Author: Mikel de Elguezabal Méndez, born in Montpellier, 6th May 1977. Biologist, University Professor in Universidad de Oriente, Cumaná, Venezuela, since 2004 (Actually in Universidad Pública de Navarra, postgraduate's studies, Spain). Entrepreneur at A6 GMP+Food Hygiene Services. President of Fundación Luís Elguezabal Aristizabal (LEA). Father of Nerea, Julen Ignacio and Maitena and married with Yaile Acosta Núñez since 2003. Believer.

@A6Labs

Tout le Monde peut cuisiner avec les BPF's!
Anyone can cook with the GMP's!

Chiloé Curantos! Cooking with hot stones under earth, covered by great green leaves!

Kiviak is a traditional Inuit food made of fermented auks! 7 months in a Seal skin bag

Milk (water+casein+fatty acids+lactose, vitamins and minerals) come from many sources: in a clockwise direction, yak, camel, llama, moose, cow and mare!

Biodiversity +Ethnodiversity= food diversity!